Introduction

Get ready to add some interesting, innovative and beautiful design elements to your yard!

This guide will show you how to implement different types of projects, including rain gardens, rain barrels and permeable pavement. Step by step instructions and worksheets are found throughout this guide.

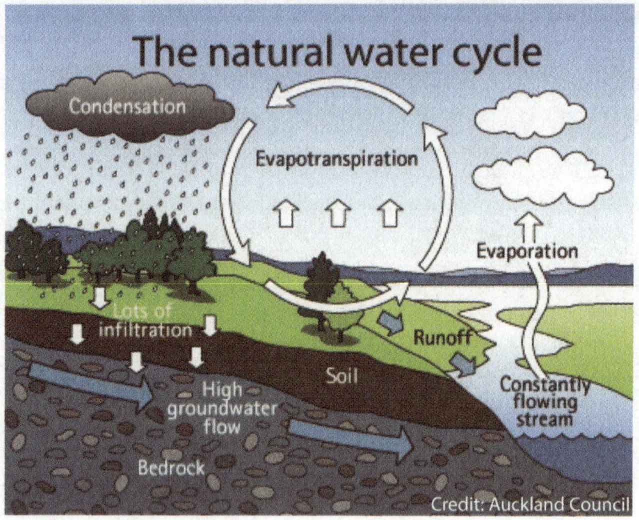

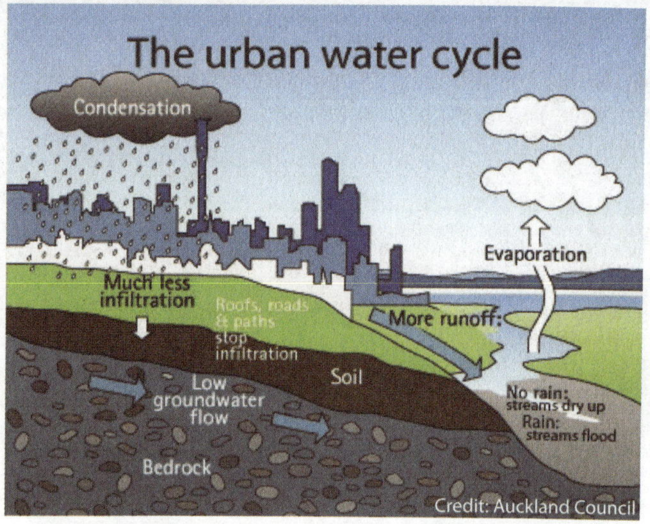

Green Your Grounds

Rain gardens, rain barrels, permeable pavement and other storm water management technologies work to conserve and clean fresh water where it lands – on our properties. To understand how it works, let's take a trip back to our Grade 5 Science class.

The Water Cycle
Remember the water cycle? That diagram showing water evaporating, condensing into clouds and falling in water droplets to represent precipitation (rain and snow). Arrows were drawn in to show the precipitation being absorbed into the ground or running off into nearby lakes and streams. This is the journey that water has been making for millions of years.

Children today draw a slightly different diagram. Where we drew a big arrow indicating that most of the precipitation was absorbed into the ground and a little arrow indicating a small amount of run off, today's children draw the opposite. In urban centers, there is very little opportunity for precipitation to be absorbed into the ground. It encounters roofs, driveways, parking lots, roads, pools, patios and compacted soils.

Now, the runoff arrow is the big arrow. When precipitation lands on paved or hard surfaces, it just runs over the land until it finds a spot where it can either be absorbed or enter the nearest water body. No big deal, right? The hydrological cycle still goes on. It's just been modified slightly. However, the increased runoff is having an effect on our water quality. We get our drinking water from our local bodies of water and runoff is creating a pollution problem.

The Pollution Problem
When water runs downhill, it tends to pick up everything in its path. Remember the slight oil leak from your car? That oil has been given a lift to your local stream during the last rainfall. How about that time you forgot to stoop and scoop when you took your pooch for a walk? Yup, that took a journey into our drinking water source too. The garbage at the curb, the fertilizer on the lawn, the soap from washing the car in the driveway – all have been given a lift to your local stream by rain water. In fact, about 70% of the pollutants that enter the Great Lakes enter them through runoff (US EPA).

引言

准备好增加一些有趣、新鲜、漂亮的设计元素到你的院子里！

这本手册将告诉你如何实施不同类型的海绵城市雨水景观项目，包括雨水花园、雨水桶和透水铺装等。在这本手册中，通过分步解说和工作表帮你顺利完成所有的工作。

绿化你的土地

雨水花园、雨水桶、透水铺装和其他雨水管理技术，在我们自己的院子中都可以使用，以保存和净化淡水资源。为了更好地理解这些设施是如何工作的，让我们一起来回顾五年级科学课程的内容。

水循环

还记得水循环吗？图 1 展示了水蒸发后成为水汽，水汽在大气中凝结变成云，随着大气运动又变成小水滴（雨或雪）降落的整个水循环过程。图中的蓝色箭头显示了降水首先由地表吸收和下渗，然后汇集成地表径流流到附近的湖或溪流当中。这是百万年以来地球上的水循环过程。

孩子们今天画的图稍有些不同。在图 1 中，蓝色大箭头表示大部分的降水被土壤吸收并下渗，蓝色小箭头代表只有少部分的降水汇集成地表径流。而今天孩子们的画正好相反。在城市的中心，由于存在着大量不透水的屋顶、机动车道、停车场、道路、水塘、露台和压实的土地等，降水很难被土壤吸收并下渗，因此如图 2 所示，大部分降雨汇集成了地表径流。

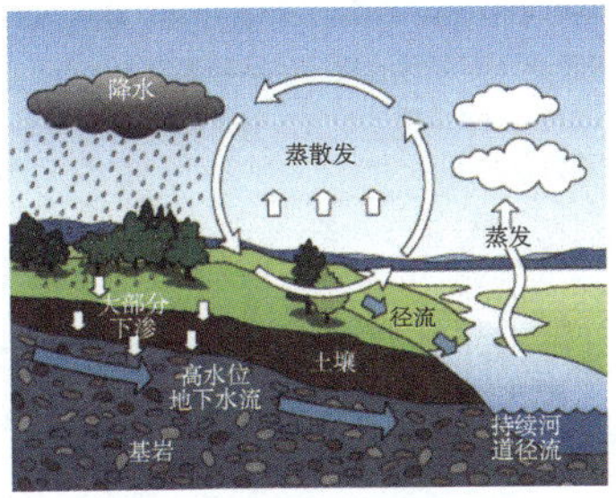

图 1 自然水循环

图片来源：奥克兰委员会

图 2 城市水循环

图片来源：奥克兰委员会

现在，径流的箭头是大箭头，代表产生了更多的地表径流。当沉淀物落在不透水铺装路面或坚硬表面时，它仅仅流经地表，直到可以被吸收或流入附近水体。没什么大不了的，对吧？水循环还在继续，它只是被改造了。然而增加的径流却对水质产生了影响。我们从当地的水体中获取饮用水，而更多的径流却造成了水体的污染问题。

污染问题

当水向下游汇集和流动时，会携带流经场地的所有污染物。还记得你的汽车轻微漏油吗？那些汽油已经通过最近的一次降雨流到了当地的河流中。还记得某次你带着宠物狗外出散步时，忘记了处理它的排泄物吗？是的，这些排泄物也会通过水的流动和携带进入到我们的饮用水源中。路边的垃圾、草地中的肥料、汽车的洗车剂等，所有的这些都将通过降雨进入到当地的水系中。数据显示，大约 70%的污染物通过河流系统最终流入了北美五大湖区（此信息来自美国环境保护局）。

专业词汇

innovative/ˈɪnəʊveɪtɪv/*adj*. （产品、观念等）革新的；创新的

element/ˈelɪmənt/*n*. （抽象事物的）基本组成部分，要素

instruction/ɪnˈstrʌkʃn/*n*. 指令，命令；说明，装配指南

worksheet/ˈwɜːkʃiːt/*n*. （学生做的）活页练习题；工作单

fresh/freʃ/*adj*. 革新的，创新的；新颖的；有创新精神的

property/ˈprɒpəti/*n*. ［总称］个人财产，资产；所有物；特性，性质，性能

diagram/ˈdaɪəgræm/*n*. 图解，简图；示意图

evaporate/ɪˈvæpəreɪt/*vt*. 使…蒸发；使…脱水；使…消失

condense/kənˈdens/*vi*. 冷凝，凝结，凝聚；*vt*. 使浓缩；使凝缩；压缩

droplet/ˈdrɒplɪt/*n*. 小滴，微滴

evapotranspiration/ɪˌvæpəʊˌtrænspɪˈreɪʃn/*n*. 蒸发蒸腾作用

evaporation/ɪˌvæpəˈreɪʃən/*n*. 蒸发；消失

infiltration/ˌɪnfɪlˈtreɪʃən/*n*. 渗透；渗透物

groundwater/ˈgraʊndˌwɔːtə(r)/*n*. 地下水（指地下土壤或岩石孔隙中的水）

bedrock/ˌbedˈrɒk/*n*. ［地质］基岩

runoff/ˈrʌnɒf/*n*. ［水文］径流；决赛；流走的东西

dry up（使）干透；（使）干涸；（使）枯竭；（水分）全干；（井等）干涸；逐渐枯萎

flood/flʌd/*n*. 洪水，水灾；眼泪的涌出，感情的迸发

constantly/ˈkɒnstəntli/*adv*. 不断地；时常地

opposite/ˈɒpəzɪt/*adj*. 对面的；对立的；完全不同的；截然相反的

encounter/ɪnˈkaʊntə(r)/*vt*. 遇到（困难或敌对的事）；面临，邂逅；意外遭遇（敌人或对手）；与…邂逅，偶然遇见

parking lots 停车场

patio/ˈpætɪəʊ/*n*. （连接房屋并铺有地面的）露台，平台；露天庭院，天井

compacted soil 紧实土；压实土；夯实土

pollution/pəˈluːʃn/*n*. 污染，污染物

downhill/ˌdaʊnˈhɪl/*adv*. 向坡下，向山下；〈喻〉走下坡路地；情况逐渐恶化

give a lift 让…搭车

stoop/stuːp/*vi*. 低头；俯身，弯腰；堕落；*n*. 弯腰，屈背

scoop/skuːp/*n*. 长柄勺，球形勺；铲斗，戽斗

pooch/puːtʃ/*n*. 〈非正式〉狗

fertilizer/ˈfɜːtəlaɪzə(r)/*n*. 肥料

pollutant/pəˈluːtənt/*n*. 污染物

Reducing Our Usable Drinking Water

Runoff is not only having an effect on our water quality, it is also having an effect on our water quantity. When it rains or the snow melts, most of the water now moves directly into local rivers and streams. On its way, it often causes severe flooding and erosion damage. Since rain water is not being absorbed into the ground, it is not recharging the groundwater system (water stored underground). This is bad news for people who get their drinking water from wells. And since groundwater slowly releases back into our rivers, streams and lakes, less groundwater means less water for our surface water systems.

So What to Do?

Decreasing the number of pollutants available for runoff to pick up on its way to the rivers is a good first step. You can also help prevent water from running off of your property in the first place. This is where your lot level stormwater management projects come in.

Roofs, driveways, parking lots, roads, and patios are all impermeable surfaces, meaning they don't allow water to absorb into them. The goal of stormwater management projects is to either change impermeable surfaces into permeable surfaces, or to direct the water that lands on one of these impermeable surfaces to an area where the water can absorb into the ground. Rain gardens, permeable pavement and the other projects outlined in this guide do just that.

Municipal Infrastructure Costs on the Rise!

The municipal infrastructure required to move water away from urban areas and store it is very expensive. As our urban areas expand, additional infrastructure is required. Not to mention the amount of maintenance required for aging infrastructure. Keeping water on our properties will reduce the need for these expensive features.

Stormwater User Fees

In several cities in the United States and now in Canada, residents are being charged a stormwater user fee. This fee is calculated by the amount of impermeable area found on their property. They have found that in an urban setting, a property's impermeable area is the greatest factor affecting stormwater quality and quantity.

Stormwater credits are often available to homeowners who implement methods to manage stormwater, such as those found in this guide. If a homeowner implements these practices on their property, they will see a reduction in their user fee.

Although your municipality may not currently participate in a stormwater user fee program, it may be soon. The City of Kitchener, just to the west of Toronto, passed a stormwater user fee program in 2011, whereby property owners are now being charged a user rate on their monthly utility bill.

WHAT IS STORMWATER?

Stormwater is water from rain or melting snow that does not soak into the ground. It runs over rooftops, paved areas and bare soil. As it flows, the stormwater picks up and transports the following pollutants:

- Sediment
- Pet waste
- Pesticides
- Fertilizer
- Automobile fluids (oil, grease, gasoline, antifreeze)
- De-icing products (road salt and fertilizers)
- Grass clippings, leaves, and other yard waste
- Cigarette butts and other litter

WHAT IS A STORM DRAIN?

A storm drain (aka storm sewer) is designed to drain excess water from driveways, streets and parking lots. In urban areas built before 1990, the water that enters a storm drain flows directly into the nearest water body with no treatment. Any pollutants that the water picks up along the way are deposited directly into your local lake or river.

In some older communities, downspouts are connected to a combined sewer system. In a combined sewer, there is only one pipe which carries both sanitary and storm drainage. During dry weather, combined sewers carry all contents to treatment plants. However, during wet weather, the amount of water may exceed the treatment plant's capacity and some of the sewage water overflows untreated into the lake.

减少可用的饮用水资源

径流不仅决定着饮用水的水质，同时也决定着饮用水的水量。当降雨或是降雪融化时，大部分的水将直接流到当地的溪流和河道中。在地表径流产生的过程中，将会产生严重的洪水灾害和土壤侵蚀。由于雨水不能通过地表土壤下渗，因此也不能有效地补给地下水（导致地下水储量减少），这对于那些从井中获取饮用水的居民来说是个坏消息。同时，由于地下水缓慢地流动从而补充江河湖泊，地下水储量的减少也意味着地表水系的水量将变少。

那么，该做些什么呢？

首先应该减少污染物的数量，从而避免地表径流形成时在流经的地方被水流携带走。然后，您能在自己家的庭院中，也就是降水最初接触的地方，阻止降雨转化成地表径流。自己家的庭院就是大量雨水管理项目的有效用武之地。

屋顶、行车道、停车场、道路和露台等都是不透水表面，这就意味着它们不能吸收和渗透降水。雨水管理项目的目的是，要么让不透水的表面变成可透水的，要么引导降水汇集到能被吸收和下渗的区域。雨水花园、透水铺装和本手册中列出的其他项目都是按这个思路来设计的。

市政设施的成本在增加！

汇集城市区域的降水并储存起来的市政设施的造价是非常昂贵的。由于城市面积不断扩大，需要增加更多的市政设施，还有大量需要维修的老旧市政设施。将降水在自家的庭院进行下渗和储存处理，将减少对造价高昂的市政设施的需求。

雨水处理使用费

美国和加拿大的一些城市向居民收取雨水处理使用费。该费用的计算依据是居民住宅和庭院的不透水区域覆盖面积的大小。在城市中，不透水区域面积是水质和水量的最主要影响因素。

对于那些实施雨水管理措施的房主来说，雨水信贷是可行的。例如，在本手册中能够实施的措施，如果房主在他们的住宅和庭院中实施这些做法，他们的雨水处理使用费将会减少。

虽然你所在的市政当局目前可能还没有实施雨水用户收费，但可能很快就会实施。位于多伦多西部的 Kitchener 市，在 2011 年通过了一项雨水用户联邦计划，该市的房屋业主每月都要缴纳雨水处理费用。

什么是雨水？

雨水是指没有渗入地下的降雨或融雪。它在屋顶、不透水铺装路面和裸露的土地上汇集和流动。它在流动时收集并携带下列污染物：

- 沉积物
- 宠物排泄物
- 农药
- 化肥
- 汽车液体（油、润滑油、汽油、防冻剂）
- 除冰产品（道路盐和肥料）
- 草屑、树叶和其他庭院垃圾
- 烟头和其他垃圾

什么是雨水管道？

雨水管道（又称暴雨排水道）的设计是为了从车道、街道和停车场排出多余的水。在 1990 年以前建造的城区，进入雨水管道的水直接流入最近的水体，没有经过任何处理。所有的污染物都直接沉积到当地的湖泊或河流中。

一些较老旧的社区没有进行雨污分流，房屋的落水管通常直接与市政综合排污管道相连通。在排污系统中只有一个管道，它不仅排放其他的污水，也排放雨水。在天气干旱时，下水道将所有的污水输送到处理厂。在多雨季节，雨水和污水的数量可能会超过处理厂的处理能力，一些没有经过处理的污水将会直接排放到湖中。

专业词汇

water quality 水质

water quantity 水量

erosion/ɪ'rəʊʒn/*n.* 侵蚀，腐蚀

recharge/ˌriː'tʃɑːdʒ/*vt.* 给（电池、电池驱动装置）再充电；再充满；再注满；再灌满；再倒满

release/rɪ'liːs/*vt.* 释放，解放；解除对…的限制；免除…的义务；解放…的力量

available/ə'veɪləbl/*adj.* 可利用的；可获得的；可供支配的

stormwater management projects 雨水管理项目

outline/'aʊtlaɪn/*n.* 轮廓图；略图，草图；提纲，大纲

municipal infrastructure 市政设施

expand/ɪk'spænd/*vt.* 扩大；扩充；扩展；发展

additional/ə'dɪʃənəl/*adj.* 另外的，附加的；额外的，更多的

maintenance/'meɪntənəns/*n.* 维持；保持；供养，抚养

user fees 使用费

resident/'rezɪdənt/*n.* 居民；住院医生；房客

homeowner/'həʊməʊnə(r)/*n.* 自己拥有住房者

reduction/rɪ'dʌkʃn/*n.* 缩减；降低；规模（或数量）被减少之物

municipality/mjuːˌnɪsɪ'pæləti/*n.* 市政当局

participate/pɑː'tɪsɪpeɪt/*vi.* 参与，参加

utility bill 账单；物业账单；费用单

soak/səʊk/*vt.* 浸透，浸湿（某物）；*vi.* 浸入（水中或其他液体）；（液体）透过，完全渗透

pave/peɪv/*vt.* 用石板（或砖）铺（路）设；铺砌；铺筑

sediment/'sedɪmənt/*n.* 沉淀物；渣滓

antifreeze/'æntifriːz/*n.*（多用于汽车散热器中的）防冻剂

de-icing *v.* 除冰；除冰带；*adj.* 除冰的

grass clippings 草屑

cigarette/ˌsɪgə'ret/*n.* 纸烟，香烟，卷烟

drain/dreɪn/*vt.* 排去…中的水（或其他液体）掉水；*vi.* 流掉，流走；流入，注入

treatment/'triːtmənt/*n.* 对待，待遇；治疗，医疗；疗程，处理

deposit/dɪ'pɒzɪt/*v.* 存放；储存；放置

community/kə'mjuːnəti/*n.* 社会；社区；共同体；团体

downspout/'daʊnspaʊt/*n.* 落水管（将雨水从屋顶排至水沟）；水落管

sewer system 污水管道系统

sanitary/'sænətri/*adj.*（尤指与提供污水排放设施及清洁饮用水有关的）公共卫生方面的；公共保健的；卫生的；清洁的

overflow/ˌəʊvə'fləʊ/*vi.*（尤指液体）溢出；泛滥；充溢；*n.* 充满，洋溢；*vt.* 淹没（某一地表，地面）

sewage/'suːɪdʒ/*n.* 污水；下水道；污物

Site Assessment

Any successful landscaping project begins with a carefully thought-out plan. Before starting any of the techniques presented in this guide, complete the following:

- *Sketch the physical characteristics of your property. (page 8)*
- *Calculate the slope of your property. (page 12)*
- *Determine the type of soils present. (page 16)*
- *Examine the soil infiltration rate (how long it takes for water to soak into the ground). (page 14)*
- *Verify the height of the seasonal water table. (page 16)*

WORKSHEET

Sketch your property, including all buildings (house, shed, garage), downspout locations, grassed areas, hard surfaces (driveways, walkways, patios) and existing gardens, decks and pools. Disregard the 'My Site Analysis' box, you will fill this in later.

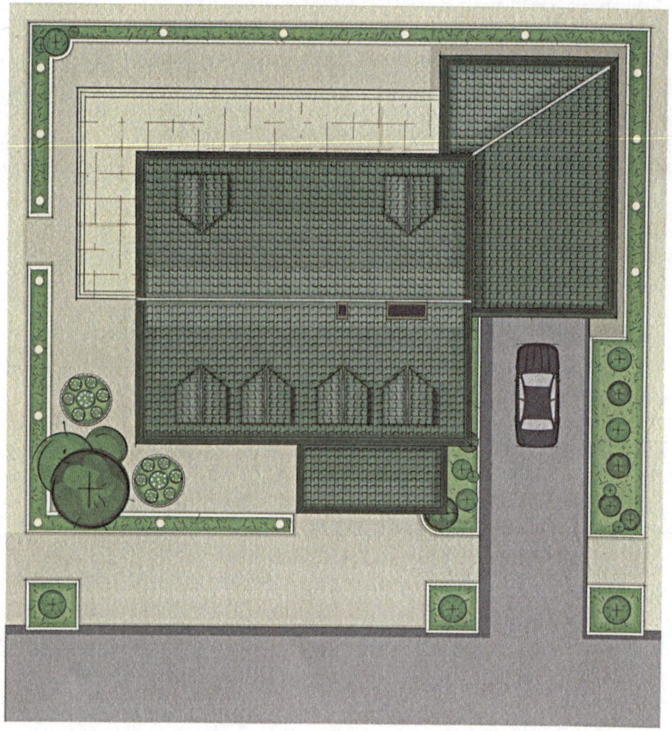

场地评估

任何一项成功的绿化工程都是从一个深思熟虑的计划开始的。在开始本手册中所涉及的技术之前，请完成以下内容：

- 画出住宅和庭院的物理特征。（第 9 页）
- 计算庭院的坡度。（第 13 页）
- 确定当前的土壤类型。（第 17 页）
- 检查土壤的渗透率（水渗入地下的时间）。（第 15 页）
- 核实季节水位高度。（第 17 页）

工作表

画出住宅及庭院的简图，包括所有的建筑物（房屋、棚、车库），落水管的位置、草地、硬化的地面（车道、人行道、天井）和现有的花园、平台和水池。先忽略“场地分析”计算框，随后再填这些内容。

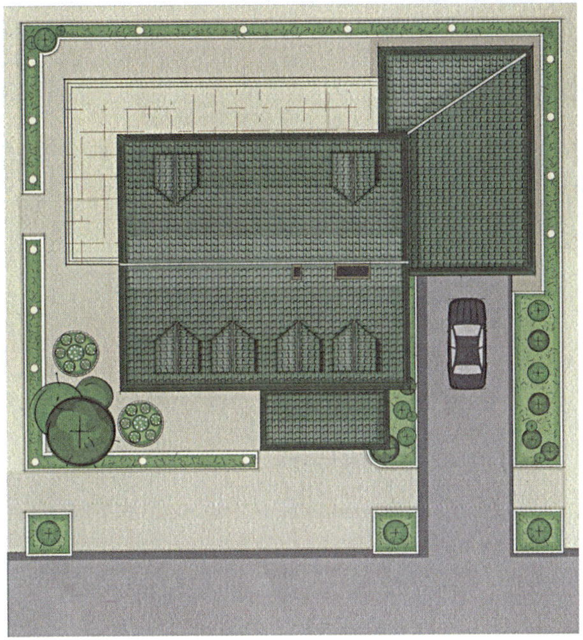

专业词汇

thought-out plan 深思熟虑的计划

slope/sləup/*n.* 斜坡；坡地；倾斜；斜度；度坡；斜率

calculate/ˈkælkjuleɪt/*vt.* 计算，核算（某物数额）；认定，以为；猜测，判断；*vi.* 预期，指望；依靠

verity/ˈverəti/*n.*（极其重要的）真原则，真信仰；事实；真理

height/haɪt/*n.* 高度，身高；高地，高处；极点，顶点

sketch/sketʃ/*n.* 素描，速写，草图；短剧；独幕剧

deck/dek/*n.* 甲板；舱位；层面；层板；（音响设备的）放音装置，录音装置

disregard/ˌdɪsrɪˈgɑːd/*vt.* 不顾；忽视

My Site Analysis

Slope:

_____ cm (rise) / ______ cm (run) = _______cm

Soil type (circle one):

Loam Sandy Clay

Infiltration rate (circle one):

Less than 2 hrs More than 2 hrs

场地分析

坡度：

______ cm（高度）/______ cm（水平长度）=________

土壤类型（在相应的选项画圈）：

壤土　　砂土　　黏土

渗透率（在相应的选项画圈）：

少于 2 小时　　多于 2 小时

专业词汇

loam/ləʊm/*n* 肥土；壤土

sandy/ˈsændi/*adj*. 覆盖着沙的；含沙的；多沙的

clay/kleɪ/*n*. ［土壤］黏土；泥土；陶土；似黏土的东西

Slope

Slope is important in stormwater management because it defines how water flows over your property. Knowing where the water flows will help you in determining the location and type of stormwater management project that is best for you.

Home Exercise

An easy way to find out how water flows on your property is to pour a full bucket of water and watch where it goes. Be sure to pour the water at the base of each downspout. This will also tell you if the slope of your property is directed away or toward your house.

> **REMEMBER**
>
> Once you know the direction of water flow from each of your downspouts, refer back to your property sketch. Draw arrows from each downspout showing where the water flowed.

Sloping Away From House

Ground that slopes away from the house is considered a 'good slope' because any water in this area will generally run away from the foundation of the house. This decreases the chance of water coming into contact with the foundation below ground and entering the basement.

Sloping Towards House

Ground that slopes toward the house is generally considered a 'bad slope' as any water that enters this area will run down the foundation of the house and increase the chance of water entering the basement. Any areas of your property that have a slope toward the house should be fixed immediately to prevent water from coming into contact with your foundation. Consult a landscape contractor if you find that you have a slope towards your house and don't know how to fix it.

倾斜方向

倾斜方向在雨水管理中是非常重要的一个因素，它决定着雨水将如何流经你的院子。了解雨水从哪里流过有助于决定采取最佳的雨水管理措施。

家庭练习

确定雨水在院子中是如何流动的最简便的方法是倒一桶水在院子里，然后观察其流动的过程。要确保在每根落水管中都倒入足量的水，这将直观地告诉你院子的倾斜方向是会导致雨水直接排走还是流向你的房子。

提示

一旦知道院子中每根落水管的水流方向，就可以返回到住宅及庭院简图中，画上箭头标示出每根落水管出水的流动方向和流经过程。

正确的倾斜方向

正确的倾斜方向使水快速排走

斜离房屋的地面被认为是“较好的斜坡”，因为这个区域的任何降水都会从房子的地基周边迅速地排走。这会大大降低水与地基接触、流入地下室的机会。

错误的倾斜方向

错误的倾斜方向使水流向房屋

斜向房屋的地面通常被认为是“不利的斜坡”，因为进入这一区域的水都将沿着房屋的地基向下流动，增加进入地下室的可能性。如果你的房子存在着斜向房屋的地面，应该立即修整，以防止水与地基接触。如果发现你的房子有不利的斜坡，又不知道如何修理，请咨询一下相关专业公司。

专业词汇

define/dɪ'faɪn/*vt.* 下定义，明确说明（或解释）；标出界线或范围

location/ləʊ'keɪʃn/*n.* 地点；设置；安置；位置；地址；外景拍摄场地

pour/pɔːr/*vi.*（尤指液体）涌流；*vt.* 倾倒，倒（液体），灌，注；倒；倾泻；倾吐

bucket/'bʌkɪt/*n.* 桶，水桶，吊桶

refer/rɪ'fɜː(r)/*vi.* 提到，淡到；指；参考；*vt.* 提交做出决定，进行处理

arrow/'ærəʊ/*n.* 箭，箭头符号（用以表示方向或位置）

foundation/faʊn'deɪʃn/*n.* 地基；基础；基本原理；创办（机构，组织）

decrease/dɪ'kriːs/*vi.* 减少，减小，减弱，降低；*vt.* 使减少，使减小，使减弱，使降低

contact/'kɒntækt/*n.* 接触；见面；联系；交往；关系

basement/'beɪsmənt/*n.* 地下室；［地质］基地；基岩，基底杂岩

increase/ɪn'kriːs/*v.* 增长；增加；增进；增强提高

prevent/prɪ'vent/*vt.* 妨碍，制止；阻止；阻挡

WORKSHEET
Measuring Slope

The following instructions will help you determine the slope under each of your home's downspouts. Slope can vary from one downspout to the next so you need to look at each one individually. Once you have calculated each slope, be sure to write the numbers on your property sketch on page 8.

Tools Required:

- Carpenter's level
- Metre stick
- Straight wooden board (a piece of 2x4, approximately 1.5 metres long works best)

Steps:

1. Place the wooden board along the slope you want to measure.
2. Place the level on top of the wooden board, and lift the lower end of the board until it is level.
3. Measure the distance from the ground to the bottom edge of the wooden board. This is your 'rise'. (If you measure it in centimetres, be sure to convert to metres.)
4. Measure the length of the wooden board you used. This is your 'run'.
5. Divide the 'rise' by the 'run' to obtain the slope. (For example, if the rise is 10 centimetres or 0.1metres, and the run is 2.0 metres, the slope is 0.1 metres / 2.0 metres = 0.05.)
6. Then multiply the answer by 100% to get your percentage slope. 0.05 x 100% = 5%.

How to Measure Slope

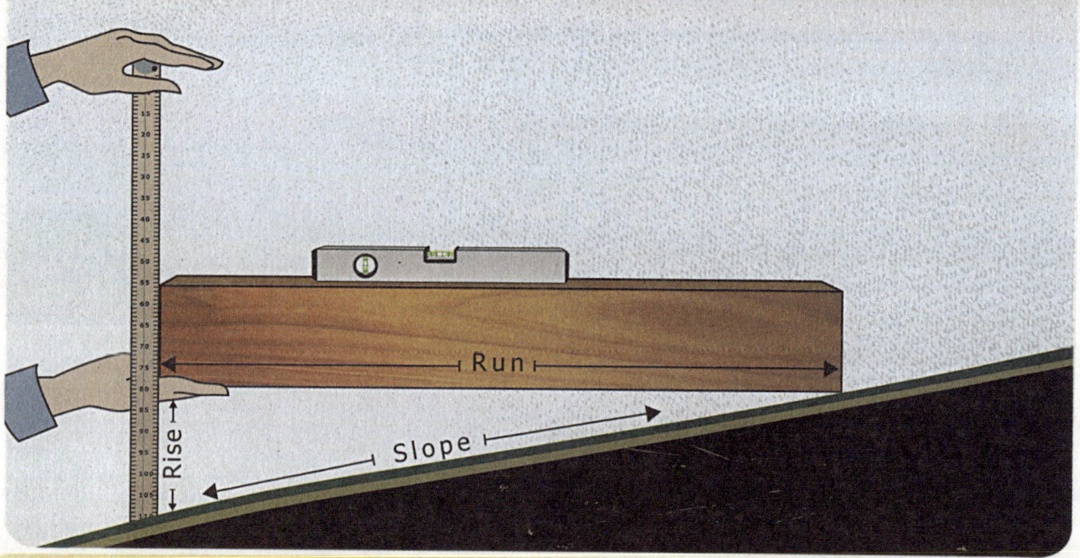

My Slope Calculation

Distance from bottom edge of wooden board to the ground (rise) = (A1)__________ cm

Convert rise into m: (A1)__________cm / 100 = (A)__________ m

Length of board = (B) __________ m

Calculate your slope: (A)__________m / (B)__________m = (C)__________

Calculate percentage slope: (C)__________ x 100% = __________%

What Does My Slope Percentage Mean?

In general, if you have a slope between 1% – 5%, that is considered good. Once the slope goes over 5%, the water flows faster and makes it harder to soak into the ground. Slopes over 5% are not ideal for stormwater management projects presented in this guide.

REMEMBER

Once you know the slope under each of your downspouts, refer back to your property sketch. Write the slope calculation under the appropiate downspout in your drawing.

工作表

测量坡度

下面的说明将有助于确定院子中每根落水管所在地面的坡度。每根落水管下的地面坡度都不相同，所以需要单独测量。一旦计算出了坡度，一定要在第 9 页场地草图的相应位置做好标注，并写下数字。

所需工具：

- 木匠的水平尺
- 米尺
- 直木板（一块断面 2cm×4cm，大约 1.5 米长的木板，效果最好）

实施步骤：

1. 把木板放在你想测量的斜坡上。
2. 把水平尺放在木板上，把木板的下端抬起来，直至水平。
3. 测量从地面到木板的底部边缘的距离。这是“抬升高度”（如果是厘米一定要换算成米）。
4. 测量你使用的木板的长度，这是“水平距离”。
5. 用“水平距离”除以“抬升高度”得到坡度。（例如，如果抬升高度是 10 厘米或 0.1 米，水平距离是 2.0 米，坡度是 0.1 米/2 米＝0.05。）
6. 然后把答案乘以 100%，得到百分比坡度。0.05×100%＝5%。

地面坡度意味着什么？

总的来说，庭院地面的坡度在 1%～5%是比较合适的。地面坡度超过 5%，雨水会快速流动，导致很难往地下渗。坡度超过 5%的地面将不适用本手册中介绍的雨水管理项目。

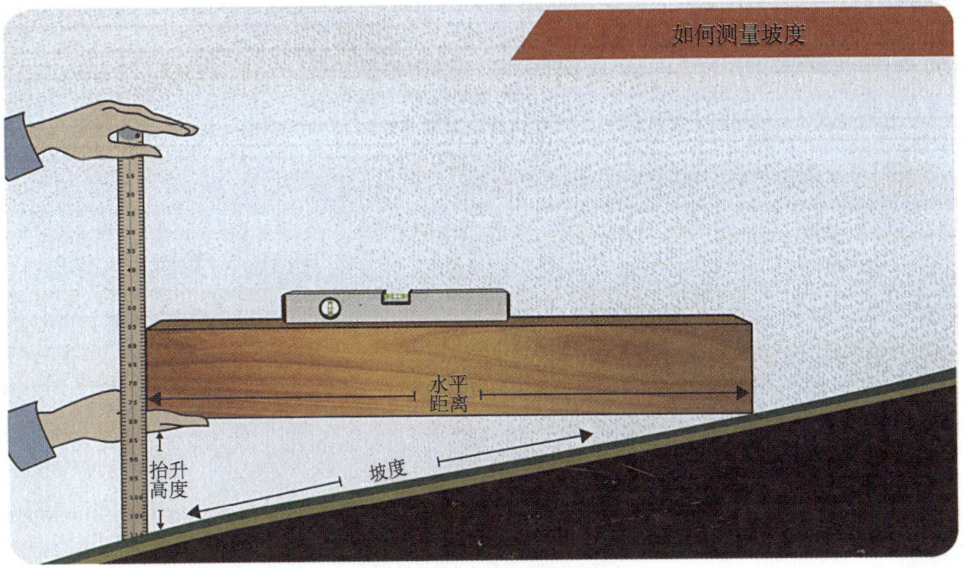

坡度计算

从木板的底边到地面的距离（抬升高度）＝（A1）________ cm

转换为 m：（A1）________ cm/100＝（A）________ m

木板水平距离长度＝（B）________ m

计算坡度：（A）________ m/（B）________ m＝（C）________

计算百分比坡度：（C）________ ×100%＝________%

提示

一旦知道了每一根落水管所在地面的坡度，就可以返回到场地草图中在落水管下面标注上。

专业词汇

carpenter/ˈkɑːpəntə(r)/*n.* 木匠

metre/ˈmiːtə(r)/*n.* 米；公尺

approximately/əˈprɒksɪmətli/*adv.* 大约，近似地；近于

centimetre/ˈsentɪmiːtə(r)/*n.* 厘米；公分

convert/kənˈvɜːt/*vt.* 使（形态、性质、功能）转变；改变；*vi.* 转变，改变

divide/dɪˈvaɪd/*vt.*（使）分开，（使）分隔；*v.* 不同意；使用分歧；[数] 除

multiply/ˈmʌltɪplaɪ/*vt.* 乘；使相乘；*vi.* 繁殖；增殖

percentage/pəˈsentɪdʒ/*n.* 百分比；百分率，百分数

Soils

In general, soils can be classified as clay, sand or loam. Clay is nutrient rich, but slow draining. Sand is quick draining, but has trouble retaining nutrients and moisture. Loam is generally considered to be the ideal gardening soil because it retains moisture and nutrients, but doesn't stay wet for extended periods.

In areas where sand content is high, infiltration of water will be fast and a great medium for almost any of the projects in this guide. However, in areas with an abundance of clay, infiltration of water will be slow. As a result, when completing many of the projects in this guide, you may need to amend (dig out and replace) the soil to allow for more water to be absorbed.

In excellent draining soils, a hole filled with water will take less than two hours to fully drain. However, as a rule of thumb, as long as the water is fully drained after 24 hours, your soils should be satisfactory for the stormwater management projects in this guide. If you find you still have water after 24 hours, chances are you have either heavy clay soils or a high water table (read below on determining water table) and should not install any of the projects in this guide.

WORKSHEET
Calculating Soil Infiltration Rate

To determine the soil infiltration rate, return to one of the downspouts you just calculated the slope for. Since this will be a likely location for one of the projects in this guide, it will give you a good idea of how fast water soaks into the ground. Complete the following steps.

Tools Required:

- Shovel
- Bucket or water hose
- Metre stick or ruler at least 30 centimetres long

1

2

5

Steps:

1. Measure 2 metres out from your home's foundation and dig a hole approximately 30 centimetres wide and 50 centimetres deep.
2. Fill the hole with water, either from a bucket or hose.
3. Let the hole drain naturally.
4. Once the water has drained, fill the hole with water again.
5. Using your ruler, measure the depth of the water. This is your initial height of water.
6. Return to the hole one hour later and measure how much the water has gone down.
7. Fill in the soil infiltration rate calculation below.

Soil Infiltration Rate Calculation

Initial height of water = (A)______cm

Height of water 1 hour later = (B)______cm

Soil infiltration rate: (A)_______cm - (B)______cm = C)______cm

Convert to millimetres: (C)________cm x 10 = (D)_______mm

Now when you divided by 1 hour, this gives you your soil infiltration rate.

(D)________mm ÷ 1 hour = ________mm/hour

REMEMBER

Now that you have determined your soil infiltration rate, be sure to add it to your property sketch details on the bottom of page 8.

土壤

一般来说，土壤可以分为黏土、砂土和壤土。黏土营养丰富但排水缓慢。砂土排水很快但难以保持营养和水分。壤土通常被认为是理想的园艺土壤，因为它既能保持水分和营养，又具有良好的排水性。

砂土含量高的区域不仅可以让水快速渗透，也是手册中所有项目中一种很好的渗透介质。然而，在有大量黏土的地区，水的渗入将是缓慢的。因此，在完成本手册中的许多项目时，可能需要改善（挖出并替换）土壤，以利于更多的水被土壤吸收。

在院子里提前挖好土坑，里面灌满水，在排水良好的土壤中不到两个小时水可以完全排干。根据经验，只要水在 24 小时内完全排干，这种土壤就应该适用本手册中的雨水管理项目。如果发现 24 小时后仍然有水，这很有可能是因为黏土土壤含量很高或地下水位较高（请阅读下面关于确定地下水位的内容），也就不适用本手册中的任何项目。

工作表

计算土壤渗透率

为了确定土壤渗透率，返回到刚刚计算出的落水管坡度。因为这将是本手册中某个项目可能设置的位置，它将使你对雨水渗入地下的速度有很好的了解。完成以下步骤。

所需工具：

- 铁铲
- 桶或软管
- 米尺或至少 30 厘米长的尺子

实施步骤：

1. 离房屋的地基 2 米之外，挖一个大约 30 厘米宽、50 厘米深的土坑。
2. 用水桶或软管将坑中填满水。
3. 让这个土坑中的水自然地下渗。
4. 一旦水排完后，再把水填满。
5. 用尺子量一下水的初始深度。
6. 1 小时后再测量水的下落深度。
7. 填写到下面的土壤渗透率计算公式中。

土壤渗透率的计算

水的初始深度＝（A）________ cm

1 小时后水的深度＝（B）________ cm

土壤渗透率：（A）________ cm －（B）________ cm＝（C）________ cm

转换为毫米：（C）________ cm×10 ＝（D）________ mm

除以 1 小时，就得到了土壤的渗透率。

（D）________ mm÷1 小时＝________ mm/h

1

2

5

提示

现在你已经确定了土壤渗透率，一定要把它添加到第 9 页下面的住宅和庭院草图上。

专业词汇

nutrient/ˈnjuːtrɪənt/*n*. 营养（物质），滋养物，养料

retain/rɪˈteɪn/*vt*. 保持；保留；

moisture/ˈmɔɪstʃə(r)/*n*. 水汽；潮气，湿气

ideal/aɪˈdiːəl/*adj*. 理想的，完美的，称心如意的，最佳的；空想的，幻想的，虚构的；不切实际的

medium/ˈmiːdiəm/*n*. 途径，方法，手段；媒质，媒介物；传导体；*adj*. 中间的，中等的；半生熟的

abundance/əˈbʌndəns/*n*. 大量，多，丰富，充足

amend/əˈmend/*vt*. 修正，改正

replace/rɪˈpleɪs/*vt*. 代替，取代；放回，把…放回原处

excellent/ˈeksələnt/*adj*. 极好的；优秀的；卓越的；杰出的

as a rule of thumb 根据经验；一般说来

satisfactory/ˌsætɪsˈfæktəri/*adj*. 令人满意的；可接受的

infiltration rate 渗滤速率；渗透速率；下渗率，渗润率

shovel/ˈʃʌvl/*n*. 铁锹，铲

water hose 水管，输水胶管；水龙带

For a more accurate analysis of your soil infiltration rate, refer to the chart below:

Soil Type	Basic Infiltration Rate (mm/hour)	Type of Soils	Drainage Time (hrs)
Sand	30+	Excellent draining soils	Less than 2
Sandy loam	20-30	Very good draining soils	2 – 6
Loam	10-20	Good draining soils	6 – 10
Clay loam	5-10	Satisfactory draining soils	10 – 16
Clay	1-5	Poor draining soils	16 - 24

Determining the Height of Your Seasonal Water Table

Every year, usually when the snow melts in spring, the water in our streams, rivers and, even underground, rises to its highest point. The water underground, called groundwater, can rise very close to the surface during this time of year. If the groundwater rises to within 3 metres of the surface, then we recommend you not continue with any of the stormwater management projects in this guide.

For assistance with determining the height of your seasonal water table, please contact:

Toronto and Region Conservation
Manager - Geoenvironmental Engineer
416-661-6600 ext.5369

or, if you live in the Region of Peel:

Region of Peel
Technical Specialist - Water Quality and Compliance
905-791-7800 ext.4646

AMENDING YOUR SOILS

If your property has clay (poor draining) soils, you may have to amend them before installing the projects in this guide. Amending your soils means replacing some of that clay with a loose soil mix using dark organic matter or a sand-compost mix to help increase your infiltration rate.

See below for guidance on how deep you should dig down when amending your soil.

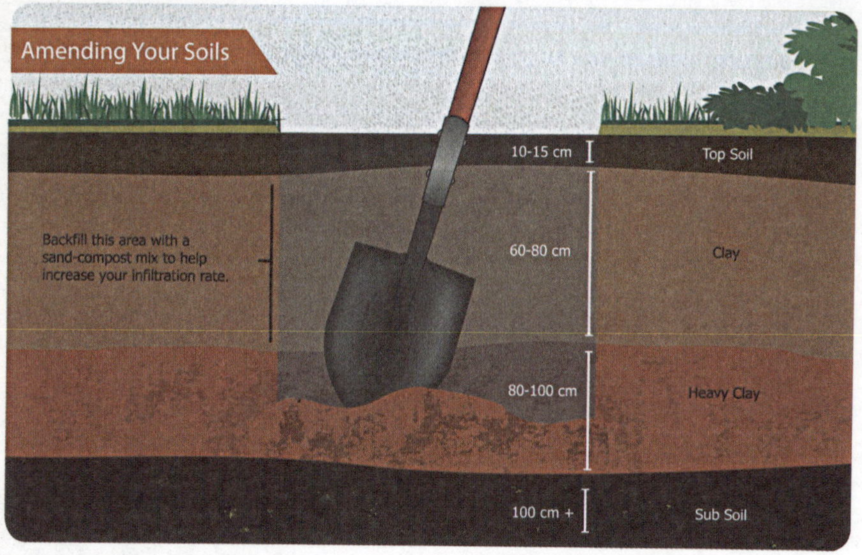

为了更准确地分析土壤的渗透率，请参考下表：

土壤类型	基本渗透速度（mm/h）
砂子	30＋
砂壤土	20～30
壤土	10～20
黏壤土	5～10
黏土	1～5
土壤类别	**排水时间（h）**
优质的排水土壤	少于 2
很好的排水土壤	2～6
良好的排水土壤	6～10
令人满意的排水土壤	10～16
排水不良土壤	16～24

确定你所在地区的季节性地下水位

通常在每年在春天雪融化的时候，溪流、河流甚至是地下水都上升到最高点。地表以下的水被称为地下水，它在每年的这个时候可能会非常接近地表。如果地下水的水位距离地表面 3 米以内，建议你不要继续使用本手册中的任何雨水管理项目。

如需协助确定季节性地下水位，请联系：

多伦多及地区保护协会

经理-地球环境工程师

416－661－6600 分机 5369

如果你生活在皮尔地区

皮尔区域

技术专家-水质和监察

905－791－7800 分机 4646

改善土壤

如果土壤中有黏土（排水不良），可能需要你在使用本手册的项目之前改善它们。改善土壤意味着用一种松散的混合土来代替一些黏土，使用深色的有机物质或砂土混合料来增加土壤的渗透率。

请参阅下图，了解改善和替换土壤时应开挖的深度。

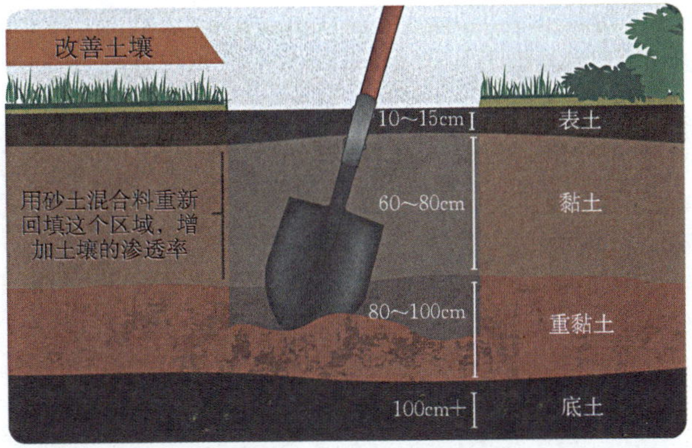

专业词汇

water table 地下水位

melt/melt/*vi.* 融化；熔化；*vt.* 使融化；使熔化；使软化；使感动

groundwater/ˈgraʊndˌwɔːtə(r)/*n.* 地下水

recommend/ˌrekəˈmend/*vt.* 推荐，举荐；劝告；建议（某人）做某事；使受欢迎

install/ɪnˈstɔːl/*vt.* 安装；设置（设备，机器）；使就任，使就职

mix/mɪks/*vt.* 使混合，拌和；*vi.* 参与；相混合；交往；*n.* 混合；混合物；混乱

organic/ɔːˈgænɪk/*adj.* （与）生物体（有关）的；（与）有机体（有关）的；构成整体所必需的，不可分割的，有机的

compost/ˈkɒmpɒst/*n.* 堆肥；混合物

Naturalization: Greening Your Grounds with Plants

What is Naturalization?

Naturalization is about creating dynamic, natural landscapes; with less lawn, and more trees, shrubs and flowers. Naturalized gardens feature a wide variety of native plants, complementing and enhancing any style of home and garden. Naturalized yards work in balance with existing natural conditions (soils, climate, water balance), rather than working against them. This results in a low-maintenance landscape that manages stormwater on your property while providing habitat for wildlife, birds and butterflies.

Like any type of landscaping, there are many different ways to bring naturalization to your property. You can choose to naturalize a small corner of your yard, a large area of your property, or you can introduce naturalization in pockets throughout your existing lawn. Naturalized gardens can range from yards full of perennials to dry river beds with a sprinkling of plants. They can include meadows of wildflowers or terraced patios. The options are endless!

Photo Credit: Paul G. O'Hara, Blue Oak Native Landscaping

Photo Credit: Paul G. O'Hara, Blue Oak Native Landscaping

Photo Credit: Paul G. O'Hara, Blue Oak Native Landscaping

Benefits of Naturalization:

- Reduces your energy costs by providing summer shading and winter windbreaks.
- Absorbs more water than a grassed area during a rain storm.
- Provides habitat for birds, butterflies and other wildlife.
- Improves the air quality around your home by increasing oxygen production while absorbing harmful air pollutants.
- Requires less energy, water and maintenance than a conventional lawn.
- Known to increase overall property value.

How Does Naturalization Help Manage Stormwater on my Property?

Naturalized areas are documented to absorb significantly more rainwater than a regular patch of lawn and/or impermeable surface

IS NATURALIZING MY PROPERTY RIGHT FOR ME?

	YES	NO
Do you currently have an open space of grass?	☐	☐
Would you like to cut your grass less frequently?	☐	☐
Do you want more birds and wildlife visiting your yard?	☐	☐
Is there a source of water to feed your potential garden space, such as a downspout, a rain barrel?	☐	☐
Do you want to increase the amount of shade in the summer and have a wind break for your house in the winter?	☐	☐
Would you like to enhance the look of your garden with various levels and colours?	☐	☐

If you answered 'YES' to all of the above, you are ready to begin naturalizing your property! Read on to learn how.

移植：土地绿化

什么是移植？

移植能够创造有活力的、自然的风景，需要尽量少用草坪，更多采用树、灌木和花。自然化的花园有各种各样的本地植物，来补充和美化任何风格的家庭和花园。自然化的庭院与现有自然条件（土壤、气候、水资源）相平衡，而不是与之对抗。这将使你的雨水管理项目成为低维护成本景观，同时为野生动物、鸟类和蝴蝶提供栖息地。

就像任何景观美化项目一样，有很多不同的方法可以将植物移植到庭院中。可以选择院子里的一个小角落，也可以选择一个较大的区域进行植被移植，甚至可以在现有的草坪上进行移植。自然化的花园选择范围很广泛，从满是多年生植物的院子到被少量植物点缀的干涸河床，从布满野花的草地到阶梯状的露台，选项是多种多样的！

移植的优点：

- 通过提供夏季遮阳和冬季防风来降低能源消耗。
- 在暴雨中吸收下渗的水量超过草坪。
- 为鸟类、蝴蝶和其他野生动物提供栖息地。
- 通过增加氧气的生产和吸收有害的空气污染物来改善家庭周围的空气质量。
- 比传统的草坪需要更少的能量、水和维护成本。
- 众所周知，可以提升庭院整体价值。

蓝橡树原生景观　　图片来源：保罗 G. 奥哈拉

移植植物是正确的选择吗？

你现在有一片开阔的草地吗？　是☐ 否☐

你想减少修剪草坪的次数吗？　是☐ 否☐

你想要更多的鸟类和野生动物来光顾你的院子吗？　是☐ 否☐

你的潜在花园空间是否有水源，例如落水管或雨水桶？　是☐ 否☐

你想为你的房屋夏天遮荫、冬天防风吗？　是☐ 否☐

你想要用不同的层次和颜色来美化你的花园吗？　是☐ 否☐

对于上述所有问题，如果你都回答“是”，那么你已经准备好移植植物了！继续往下阅读，学习具体做法。

专业词汇

naturalization/ˌnætʃərəlaɪˈzeɪʃn/*n*. 归化，移入；移植

dynamic/daɪˈnæmɪk/*adj*. （过程，系统）动态的；不断变化的；推动的，动力的；［物理］（与）动力（有关）的；

naturalized/ˈnætʃrəlaɪzd/*adj*. 归化的；入籍的；［动物］自然化的；驯化的；*v*. 使自然化；接受…入国籍；采纳（naturalize 的过去式和过去分词）

variety/vəˈraɪəti/*n*. 变化，多样性；品种；杂耍；［生］植物品种

complement/ˈkɒmplɪmənt/*n*. 补强，增强；足数；补足物；补足量；补语

enhance/ɪnˈhɑːns/*vt*. 增加；提高…的质量（或价值，程度）

yard/jɑːd/*n*. 围起的空地，院子，庭院；码（长度单位，等于 3 英尺，即 0.9144 米）；帆桁；横杆

low-maintenance 少量维护保养；维修费用低；工作量不大的技术维护

habitat/ˈhæbɪtæt/*n*. （动植物等的）生（长环）境；栖息地

wildlife/ˈwaɪldlaɪf/*n*. ［总称］野生动物；（某一地区的）动物群；（有时也指）植物群

pocket/ˈpɒkɪt/*n*. 衣袋；袋子；兜；小块

lawn/lɔːn/*n*. 草地；草坪

perennial/pəˈrenɪəl/*adj*. 长期的；永久的；持久的

sprinkling/ˈsprɪŋklɪŋ/*n*. 少量，稀稀拉拉的几个

meadow/ˈmedəʊ/*n*. 草地，草场，牧场

terraced/ˈterəst/*adj*. （土地）呈梯田状的，台阶形的

endless/ˈendlɪs/*adj*. 无穷尽的，无止境的；无限的

reduce/rɪˈdjuːs/*vt*. 减少（数量），降低（程度），缩小（规模）；使降至；使改变形式，使还原

energy cost 能量价格，能量消耗

shade/ʃeɪd/*n*. 荫；阴凉处；（色彩的）深浓浅淡；灯罩

windbreak/ˈwɪndbreɪk/*n*. 防风物（如防风林、防风篱、挡风墙、风障等）

oxygen/ˈɒksɪdʒən/*n*. 氧气，氧

harmful/ˈhɑːmfl/*adj*. 有害的，能造成损害的

conventional/kənˈvenʃənl/*adj*. 传统的，习惯的

frequently/ˈfriːkwəntli/*adv*. 频繁地，经常地

potential/pəʊˈtenʃəl/*adj*. 潜能的，可能的

like a patio or driveway. On many impermeable surfaces, 95% of rainwater becomes runoff, while on a naturalized landscaped area, only 20%-25% becomes runoff.

Native plants are able to tolerate periods of drought and occasional flooding from large rain storms. The deep root systems of native trees, shrubs, and wildflowers help water soak deep into the ground, recharging local ground water. Even in poorly drained soils with high clay content, a naturalized area is better at absorbing rainwater than a patch of lawn.

Naturalization Can Be Used With All the Projects in This Guide!

Including naturalized areas in your existing landscape is the easiest way to absorb stormwater on your property. Using native plants with the projects featured in this guide will help them to function as if they were out in nature. See the Chapter "Connecting the Drops" for ideas.

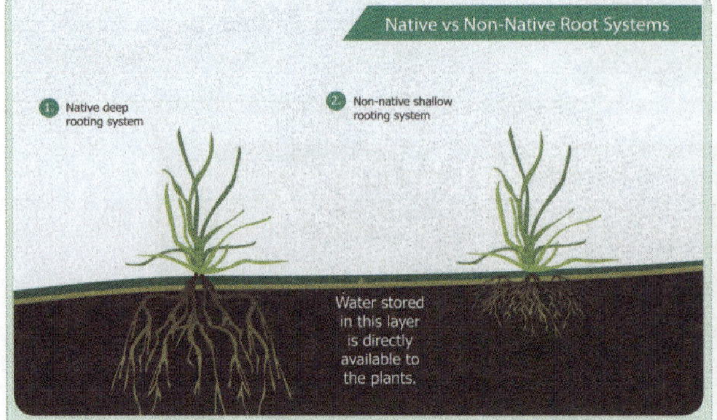

Did you know some species of native grasses and wildflowers have up to 3.5 metre root systems at maturity, compared with the 10 to 15 centimetre root systems common to most turf grasses, like Kentucky bluegrass.

WHAT ARE NATIVE PLANTS?

The soil and climate of a region in large part determine the plants and animals that naturally exist there. These species are referred to as native. By definition, a native plant is one that occurs naturally in a particular region, without human intervention. In other words, these are the plants that were growing here prior to the arrival of European settlers. Southern Ontario has its own collection of native species that have evolved together over thousands of years. Native plants are vital to creating a natural, sustainable landscape, as they provide the food and shelter needed by native insects and wildlife, and they are perfectly suited to our local growing conditions.

New England aster

Chokeberry

Butterfly weed

Wild geranium

Joe-Pye weed

New Jersey tea

庭院绿化后是如何应对暴雨的?

自然化种植的区域比常规的草坪或是不透水的表面，例如一处露台或行车道，能吸收更多的雨水。在许多不透水的表面，95%的雨水变成了径流，而在一个自然化种植的景观区，只有20%～25%的雨水变成了径流。

本地植物能够应对干旱季节和由于大暴雨造成的偶发性洪水。原生树木、灌木和野花的深层根系有助于水更快地渗入地下，有效地补充当地的地下水。即使是黏土含量高的贫瘠土壤，自然化种植的区域也比一片草坪吸收雨水的能力更强。

移植植物可以用在本手册中的任何项目中!

现有的景观包括你的自然种植区域是吸收雨水最简单的方法。在本手册中使用本地植物，将有助于它们更好地发挥作用，就像在大自然中一样。参见“雨水的连通路径”的相关章节。

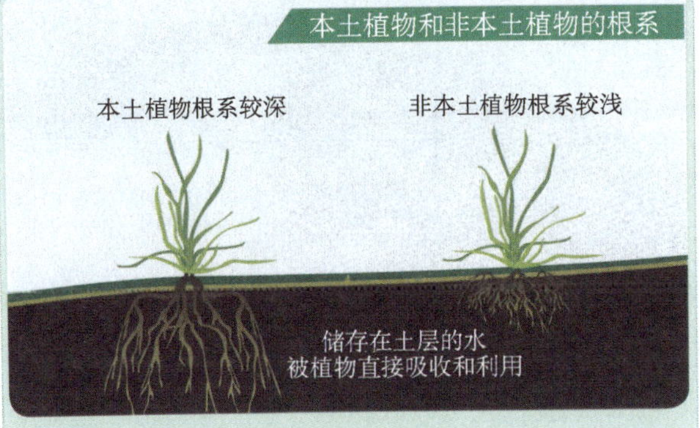

你是否知道一些本土草和野花在成熟时根系深达3.5米，相比之下大多数草类的根系只有10～15厘米深，比如肯塔基蓝草。

什么是本土植物?

一个地区的土壤和气候在很大程度上决定了自然存在的植物和动物。这些物种被称为本土物种。根据定义，本土植物就是在特定区域中没有人类干预的情况下自然生长的植物。换句话说，这些是在欧洲殖民者到来之前生长在这里的植物。南安大略有自己的本土物种，这些物种在数千年的时间里进化而来。本土植物对于创造自然的、可持续的景观至关重要，因为它们提供了本地昆虫和野生动物所需要的食物和住所，而且它们非常适应当地的环境条件。

美国紫菀

花楸果，野樱桃

蝴蝶草

野老鹳草

紫泽兰

新泽西茶

专业词汇

document/ˈdɒkjumənt/*n.* 文件

tolerate/ˈtɒləreɪt/*vt.* 容许，不干预

occasional/əˈkeɪʒənəl/*adj.* 不经常的，偶尔的

shrub/ʃrʌb/*n.* 灌木；灌木丛

determine/dɪˈtɜmɪn/*vt.* 造成；在…中起决定性作用；下定决心，决意

exist/ɪgˈzɪst/*vi.* 存在，实际上有；生存；(尤指在逆境中）生活

species/ˈspiːʃiːz/*n.* （生）物种；种；类

definition/ˌdefɪˈnɪʃn/*n.* （尤指词典中的）定义，释义；(尤指照片或屏幕图像的）清晰度

occur/əˈkɜː(r)/*vi.* 发生；出现；存在

region/ˈriːdʒən/*n.* 地区，地方，区域

intervention/ˌɪntəˈvenʃn/*n.* 插入，插手，介入；干预，干涉

European/ˌjʊərəˈpiːən/*adj.* （与）欧洲（有关）的；(与）欧洲人（有关）的

settler/ˈsetlə(r)/*n.* 移居；拓荒者

evolve/ɪˈvɒlv/*vt.* 逐步发展；演进；*vi.* 进化；演化

insect/ˈɪnsekt/*n.* 昆虫

suit/suːt/*vt.* 适合；使适应；*n.* 诉讼；恳求；套装，西装；一套外衣；*vi.* 合适；相称

Lawn Alternatives

Although we know that turf grass absorbs less rainwater than a naturalized landscape, you may want to keep some grassy areas on your property.

Most turf grasses in Ontario are a mix of Kentucky bluegrass, fescue and perennial ryegrass. These grasses require a lot of water and maintenance, and thrive in a wetter climate. A typical lawn requires up to four times as much water as other plants.

Lawn Tips:

- Let your grass go yellow through the summer drought. The grass is not dead; it has just gone into a natural dormant state. It will revive during the cooler, damper days of fall and green up again. IMPORTANT: once you have decided to let your grass go dormant, DO NOT apply enough water to green it up, then let it go back dormant. Breaking the lawn's dormancy will drain large amounts of food reserves from the plant and may kill it.
- Look for varieties of grass seed mix, which are more drought resistant and include native varieties of grass. These seed mixes are usually labeled as "Eco-grass" or "Low maintenance grass".
- Consider a clover lawn. Dutch white clover grows deep roots and needs less water to stay green than conventional grasses. If you see clover patches in lawns in the summer, they are usually green, while the rest of the lawn is brown and dormant. White clover spread throughout a lawn can make it appear green year round. White clover also resists browning from pet urine, is relatively low growing, and out-competes weeds.

If you aren't stuck on the idea of a typical turf lawn, there are many other groundcover alternatives to consider – low-growing herbs, wildflowers, mulch or decorative stone all provide beautiful, low-maintenance landscapes and are able to thrive on much less water than a typical lawn requires.

Photo Credit: Sean James, Fern Ridge Landscaping

草坪的替代品

尽管我们知道草坪吸收的雨水比自然景观植被少，但你可能仍想要在庭院保留一些草坪。

安大略的大多数草都是肯塔基蓝草、羊草和多年生的草。这些草需要大量的水资源和维护成本，在更湿润的气候中才能茁壮成长。一个典型的草坪需要的水量是其他植物的四倍。

草坪小贴士：

- 当你的草坪在干旱的夏天变黄时，其实草并没有死，只是进入了自然休眠状态。它会在凉爽、潮湿的秋天恢复，再次变绿。重要提示：一旦草坪进入休眠状态，不要试图大量浇水让它恢复绿色，让它进入休眠状态就好。打破草坪的休眠状态将会从植物中吸收大量的能量储备，并可能导致它死亡。
- 寻找各种各样的草籽混合在一起，它们会更耐旱，包括本地品种的草。这些混合不同种类的草籽通常被标记为“生态草”或“低维护草”。
- 考虑一下三叶草草坪。荷兰白色的三叶草有很深的根系，需要的水比传统的草要少。夏天，草坪上的三叶草通常是绿色的，而其余的草坪则处于休眠状态，是棕色的。白色的三叶草分布在整个草坪上，可以让它看起来一年四季常绿。白色的三叶草也会抵抗由于宠物尿液导致的褐变，而且它生长速度相对缓慢，胜过杂草。

如果你不想在院子中有草坪，还有许多其他的基础设施可供考虑——生长缓慢的草本植物、野花、覆盖物或装饰性的石头都可以提供美丽的、低维护成本的景观，并且比普通的草坪所需的水更少。

蕨类植物风景　　　　图片来源：肖恩・詹姆斯

专业词汇

alternative/ɔːlˈtɜːnətɪv/*adj.* （一个或多个事物）另外的，可供选择的；*n.* （二者或多者中）另种选择；另种可能

turf/tɜːf/*n.* 草皮；草泥，草根土；泥炭；赛马，赛马场；*vt.* 铺草皮

grassy/ˈɡrɑːsi/*adj.* （青）草的；长满草的

thrive/θraɪv/*vi.* （孩子、动物）茁壮成长，植物茂盛生长

typical/ˈtɪpɪkl/*adj.* 典型的；有代表性的

dormant/ˈdɔːmənt/*adj.* （动物）冬眠的；（似）在深睡的

revive/rɪˈvaɪv/*vi.* 复兴；复活；苏醒；恢复精神；*vt.* 使复活；使苏醒

damp/dæmp/*vt.* 使潮湿；使阻尼；使沮丧，抑制；*vi.* 减幅，阻尼；变潮湿；*n.* 潮湿，湿气；*adj.* 潮湿的

mix/mɪks/*vt.* 配制；混淆；使混合；使结交；*vi.* 参与；相混合；交往；*n.* 混合；混合物；混乱

resistant/rɪˈzɪstənt/*adj.* 抵抗的，反抗的；顽固的

Dutch/dʌtʃ/*adj.* 荷兰的；荷兰人的；荷兰语的；*n.* 荷兰人；荷兰语；*adv.* 费用平摊地；各自付账地

clover/ˈkləʊvə(r)/*n.* 三叶草；苜蓿；红花草

resist/rɪˈzɪst/*vi.* 抵抗，抗拒；忍耐；*vt.* 抵抗；忍耐，忍住；*n.* [助剂] 抗蚀剂；防染剂

urine/ˈjʊərɪn/*n.* 尿，小便

out-compete *vt.* 超过；胜出

stuck/stʌk/*v.* 刺（stick 的过去式）；*adj.* 被卡住的；不能动的

herb/hɜːb/*n.* 香草，药草；草本植物

decorative/ˈdekərətɪv/*adj.* 装饰的；装潢的

Downspout Disconnection

The rainwater that falls on the roof of a typical residential home is funneled through a series of eaves troughs, into a downspout and directed into a storm drain. Rather than collecting and using this water, we are sending it to the nearest river or stream, which contributes to flooding, erosion and water pollution. Instead of sending our rain water away, why not start using it where it lands? The first step is disconnecting your downspout.

Because the projects in this guide focus on collecting water that falls on your rooftop, we need to make sure your downspouts are not connected. If they are, disconnecting your downspout is the first project you should do before any of the other projects in this guide.

A typical house has at least four downspouts, so ensure that you are disconnecting each one, even though you may only be using one or two for a storm water management project in this guide.

Why Should I Disconnect my Downspout?

- To help re-establish the natural hydrological cycle.
- You help reduce the amount of rain water entering the storm drain system.
- Disconnecting your downspout allows water to soak into the ground, replenishing our ground water supply.
- If you collect your rain water and use it, not only are you conserving water, but you can save money on your water bill.
- By disconnecting your downspouts you can water your garden for free.

Take a look at your downspouts. Do they connect straight in to the ground (A), or do they end in an elbow piece and splash pad (B)?

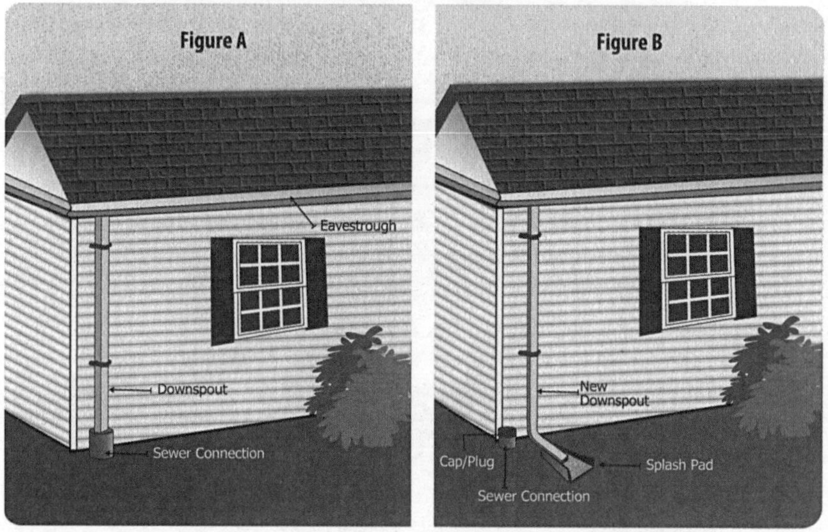

If you chose B, Congratulations! Your downspout is already disconnected. Turn to the next section to start learning how you can use the rainwater that flows out of your downspout.

If you chose A, please read on.

WHAT'S THE DIFFERENCE?

A disconnected downspout (picture B) directs rainwater that lands on your roof away from your house and onto the ground. This stormwater can be collected and re-used. A connected downspout (picture A) carries water directly from your home's downspout into the underground storm drain system. This means that 100% of the water that lands on your roof is funneled directly into the storm drain. Disconnecting your downspouts is one of the most important things you can do to help prevent stormwater pollution.

DID YOU KNOW:

Almost half of our summer water usage is dedicated to outdoor uses.

DOWNSPOUT DISCONNECTION BY-LAWS

Some municipalities have by-laws regarding downspout disconnection.

For more information, contact your local municipality.

York: 1–877–464–9675
Peel: 905–791–7800
Durham: 1–866–786–8116
Toronto: call 311

断开的落水管

雨水降落在典型的住宅屋顶后会流经一系列的屋檐集水槽，汇集到落水管中然后排入雨水沟。宁可将雨水直接排放到最近的河流或溪流中，也不愿收集和利用雨水，从而导致了洪水、水土流失和水体污染等问题的产生。为什么不从一开始在雨水降落的区域就地利用，而是将雨水排走呢？第一步要做的就是断开落水管。

本项目主要是收集降落在屋顶上的雨水，需要确保落水管是断开的。如果落水管是连通的，断开落水管是在开始本手册的其他项目之前应该做的首个项目。

一座典型的房子至少有四根落水管，即使可能只是使用其中的一根或两根来进行雨水管理，也要确保把每一根都断开。

为什么要断开落水管？

- 帮助重建自然的水文循环。
- 帮助减少进入暴雨排水系统的雨水量。
- 断开落水管可以让降水被土壤吸收并下渗，补充地下水量。
- 收集雨水并利用它，不仅可以节约用水，还可以节省水费。
- 通过断开落水管，可以免费为花园浇水。

看一下你的落水管，它们是直接连接到排水系统（图 A），还是以肘型弯管和防溅垫（图 B）结束的？

如果你选择了 B，恭喜你！你的落水管已经断开了。转到下一节开始学习如何使用从落水管流出的雨水。如果你选择了 A，请继续阅读。

有什么区别呢？

一根断开的落水管（图 B）汇集落在屋顶上的雨水，远离你的房子最终落到地面上，这样雨水就可以被收集和再利用。连接的落水管

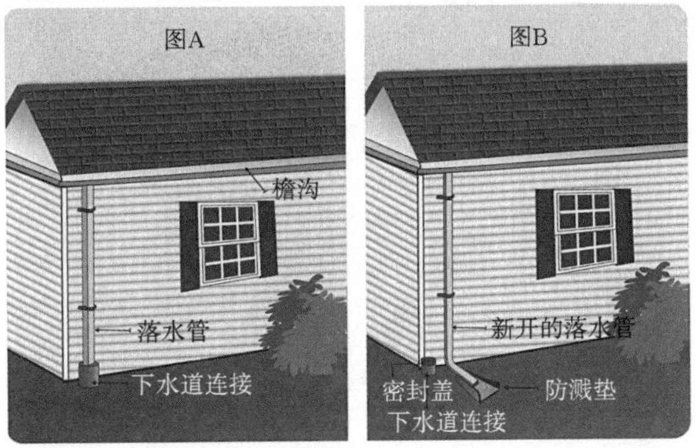

（图 A）直接将水从落水管输送到地下的雨水排放系统。这意味着落在屋顶上的水 100％都直接流入了排水沟。断开落水管是你可以做的最重要的事情之一，以防止暴雨导致的水污染。

你知道吗：

我们在夏季的用水量将近一半用于满足室外的用水需求。

断开落水管的规定

一些城市有关于“断开落水管”的规定。如需更多信息，请与当地政府联系。

约克地区：1－877－464－9675

皮尔地区：905－791－7800

杜伦地区：1－866－786－8116

多伦多地区：拨打 311

专业词汇

residential/ˌrezɪˈdenʃəl/*adj*. 作住所用的，住宅的；与居住有关的

funnel/ˈfʌnəl/*n*. 漏斗；烟囱；*vt*. 通过漏斗或烟囱等；使成漏斗形；*vi*. 通过漏斗等；成漏斗形

rather than 而不是；宁可…也不愿

contribute/kənˈtrɪbjuːt/*vt*. 捐（款）；捐助，捐献；贡献；*vi*. 贡献，出力；投稿；捐献

re-establish/ˌriːɪˈstæblɪʃ/*vt*. 重新建立，重建

hydrologic cycle 水文循环，水循环

replenish/rɪˈplenɪʃ/*vt*. 把…装满；再填满；把…重新备足，补足

collect/kəˈlekt/*vt*. 聚集，使集中；*vi*. 收集；聚集；募捐

bill/bɪl/*n*. 账单，账款；广告；账单；（金融）票据；钞票；清单；*vt*. 宣布；开账单；用海报宣传

elbow/ˈelbəʊ/*n*. 肘；肘装物；（管子的）弯曲部；L 形弯管；*vt*. 用肘击打（某人）；推挤；用手肘推开

splash/splæʃ/*vt*. 溅，泼；用…使液体飞溅；*n*. 飞溅的水；啪啪声，泼溅声；*vi*. 溅湿；溅开

difference/ˈdɪfərəns/*n*. 差异；不同；争执；分歧

re-used/ˈriːjuːzd/*adj*. 再生的；*v*. 再用；重复利用（reuse 的过去分词）

prevent/prɪˈvent/*vt*. 妨碍，制止；防止；阻止；*vi*. 妨碍，阻止

dedicate/ˈdedɪkeɪt/*vt*. 把（时间，精力）用于，致力于；献身于

by-laws *n*. 章程；附则

municipality/mjuːnɪsɪˈpæləti/*n*. 市民；市政当局；自治市或区

Getting Started

Downspout disconnection is a simple, inexpensive process that you can do yourself.

Tools Required:

- Hacksaw
- Tape measure
- Screwdriver
- Work gloves and eye protection

Materials Needed:

- Downspout elbow
- Sheet metal screws
- Downspout extension piece
- Cap or plug to seal drain pipe
- Wall bracket
- Plastic splash pad or river rocks

How Much Will my Downspout Disconnection Project Cost?

The rough cost includes:

Downspout elbow	$3
Cap or plug to seal sewer pipe	$3
Sheet metal screws (package of 12)	$2
Wall bracket/Downspout strap	$2
Downspout extension piece (10 foot length)	$12
Splash pad (optional)	$9
Total	**$31**

Steps

1. Cut the downspout approximately 23 to 30 centimetres (9″ to 12″) above the storm drain connection pipe.

2. Remove the cut piece and, if required, attach a wall bracket to secure the remaining downspout to the wall.

3. Secure a cap (or plug) onto the storm drain pipe. This is an important step as it seals the pipe, which ensures that no water, debris, or animals get into it.

4. Insert an elbow over the end of the downspout. Secure with sheet metal screws.

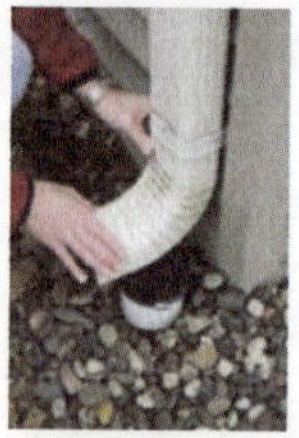

5. Add an extension piece (a length of 3 metres is recommended to provide sufficient distance from your house to prevent water from entering your basement) and secure it to the elbow with sheet metal screws. Add a plastic splash pad or a few small river rocks to protect the ground and help distribute flow.

开始操作

断开落水管是一个简单、廉价的项目，你可以自己完成。

所需工具：

- 钢锯
- 卷尺
- 螺丝刀
- 工作手套和护目镜

所需材料：

- 肘型弯管
- 金属板螺丝
- 落水管延伸段
- 排水管密封盖或密封塞
- 墙体支架
- 塑料防溅垫或卵石

落水管断开项目的费用是多少？

大致成本包括下表中的项目：［单位：加元（2019 年 1 月汇率，1 加元约等于 5.16 元人民币）］

落水管肘型弯管	3
排水管密封盖或密封塞	3
金属板螺丝（12 包）	2
墙体支架/落水管带	2
落水管延伸段（10 英尺长）	12
防溅垫（可选）	9
合计	31

实施步骤

1. 在房屋落水管上方 23～30 厘米（9～12 英寸）处切割落水管。

2. 取下切好的部分，如果需要，在墙上安装一个支架，将剩余的落水管固定在墙上。

3. 在暴雨排水管上安装一个盖子（或塞子）。这是一个重要的步骤，因为它可以密封管道，以确保没有水、碎片或动物落入管道。

4. 在落水管的末端插入一个肘型弯管，用金属板螺丝固定。

5. 增加一个延伸管（建议长度为 3 米，以提供足够的距离防止水进入地下室），并用金属板螺丝固定在肘型弯管上。添加一个塑料防溅垫或一些小的卵石来保护地面和帮助分流。

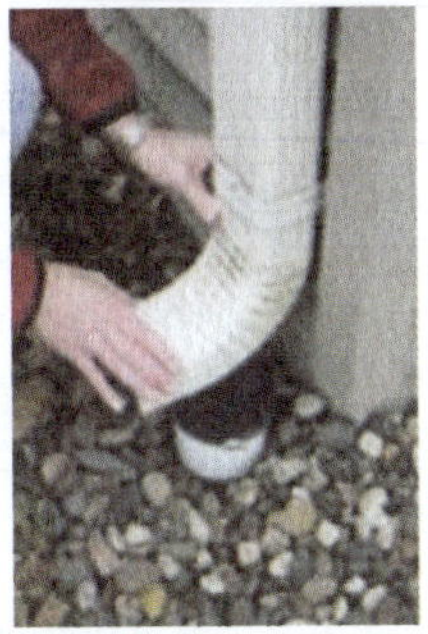

专业词汇

hacksaw/ˈhæksɔː/*n*. 弓锯，钢锯；可锯金属的弓形锯；*vt*. 用钢锯锯断

tape measure 卷尺

screwdriver/ˈskruːdraɪvə(r)/*n*. 螺丝刀，螺丝起子

glove/glʌv/*n*. 手套；*vt*. 给…戴手套

screw/skruː/*vt*. 旋，拧；用螺丝固定；*n*. 螺旋；螺丝钉；吝啬鬼；*vi*. 转动，拧

extension/ɪkˈstenʃn/*n*. 延长部分；扩大部分；附加部分；延期；扩大；伸展；电话分机

plug/plʌg/*n*. 堵塞物，插头；塞子；栓；*vi*. 塞住；用插头将与电源接通；*vt*. 插入；塞住；接插头

bracket/ˈbrækɪt/*n*. 支架；括号；墙上凸出的托架；*vt*. 括在一起；把…归入同一类；排除

plastic/ˈplæstɪk/*adj*. 塑料的；（外科）造型的；可塑的；*n*. 塑料制品；整形；可塑体

Where Should I Direct my Downspout Extension?

The downspout extension should be directed to a vegetated area at least 2–3 metres away from the foundation of your home, such as a garden, where the water can absorb into the ground. If you experience any problems with water ponding for longer than 24 hours, we recommend amending your soil (page 16). If the slope of the land away from your foundation is greater than 1%, 3 metres is not required since the water will move away from your foundation on its own (refer to page 12 for instructions on how to calculate your slope).

It is very important that your downspout extension is not directed to any areas that slope towards your house or other buildings. You want to move water away from your home, not towards it.

Where Should I NOT Direct my Downspout Extension?

Ensure that your downspout does not flow directly onto a hard surface like a driveway or walkway and make sure that the redirected rain water does not adversely impact your neighbour's property. Do not position your downspout so that it flows within 5 metres of their home or foundation.

What If I Don't Have Enough Space?

If you have limited space to extend your downspout, consider using a rain barrel (page 30), soakaway (page 64) or a rain garden (page 46) to help absorb the water the comes out of your downspout.

Maintenance

- Inspect your eaves troughs and downspouts on a regular basis. Remove any leaves or debris that are blocking the flow of water.
- Monitor the end of the downspout for pooling water. The water that exits your downspout should be absorbed within 24 hours of a rain storm. If any ponding water remains beyond that point, it is recommended that you amend your soil to help it absorb the water more quickly.(Refer to page 16 for instructions on how to do this)
- Ensure the location where water exits the downspout is aerated regularly to assist the soil in absorbing rain water.

WHAT IS AERATION?

Aeration is a simple procedure where your lawn is perforated with holes to allow water to absorb into it more easily. You can purchase an aeration tool from your local hardware or garden centre (an inexpensive one is about $30). Alternatively, you can rent a motorized aeration machine from a garden or tool rental supply store. It is best to aerate when the soil is damp. Aerating will leave plugs of soil on your lawn. They will break down on their own within a week or two.

应该在什么地方延长落水管？

落水管至少应延伸至距房屋地基 2～3 米的植被区，比如花园，在那里水可以被吸收、下渗到土壤。如果有超过 24 小时的排水问题，建议改善土壤（详见第 17 页）。如果远离地基的地面坡度超过 1%，则不需要 3 米，因为水会自然离开地基。（有关如何计算坡度的说明，请参阅第 13 页）。

很重要的一点是，落水管延伸的方向不能指向房屋或其他建筑物，正确的做法是将落水管延伸至远离房屋或其他建筑物的区域。因为你是想把水从家中引走，而不是向家中引水。

哪些地方不适合延长落水管？

确保落水管中的水不会直接流到坚硬的路面上，如车道或人行道，并确保重新定向的雨水不会对邻居的财产造成负面影响。不要将落水管放置在离家或地基 5 米以内的地方。

如果没有足够的空间怎么办？

如果安装落水管的空间有限，可以考虑使用雨水桶（详见第 31 页）、渗水坑（详见第 65 页）或者雨水花园（详见第 47 页）来帮助接收从落水管中流出的雨水。

维护

• 定期检查屋檐集水槽和落水管，清除任何阻挡水流的叶子或碎片。

• 注意检查落水管末端，防止积水。从落水管流出的水应该在暴雨后 24 小时内被吸收。如果任何积水超过了这一时间，建议改良土壤，以帮助土壤更快地吸收水分（详情请参阅第 17 页）。

• 定期检查确保落水管出水口位置通畅，以帮助土壤吸收雨水。

曝气是什么？

曝气是一个简单的过程，会让草坪有更多连通的孔隙从而更容易吸收雨水。可以从当地的五金商店或花园中心购买一个充气工具（便宜的大约 30 加元），或者从花园或工具租赁用品商店租一台曝气机。当土壤潮湿时最好通气，曝气会在草坪上留下大量的土壤，它们会在一两周内自行分解。

专业词汇

inspect/ɪn'spekt/*vt*. （多为了解情况或发现不足）检查；审视；*vi*. 进行检查；进行视察

regular/'regjələ(r)/*adj*. 定期的；有规律的；合格的；整齐的；普通的；*n*. 常客；正式队员；中坚分子；*adv*. 定期地；经常地

remove/rɪ'muːv/*vt*. 去掉，除掉，移动，迁移；开除；调动；*vi*. 移动，迁移；搬家；*n*. 移动；距离；搬家

monitor/'mɒnɪtə(r)/*n*. 监视器；监听器；监控器；显示屏；班长；*vt*. 监控

aerate/'eəreɪt/*vt*. 充气；让空气进入；使暴露于空气中

aeration/eə'reɪʃn/*n*. （环境）曝气；通风；充气

perforated/'pɜːfəreɪtɪd/*adj*. 穿孔的；有排孔的

purchase/'pɜːtʃəs/*n*. 购买；紧握；起重装置；*vt*. 购买；赢得；*vi*. 购买东西

hardware/'hɑːdweə(r)/*n*. 硬设备（指工具、机器等耐用设备）；计算机硬件；五金器具

inexpensive/ˌɪnɪk'spensɪv/*adj*. 花费不多的，便宜的

alternatively/ɔːl'tɜːnətɪvli/*adv*. 或者，作为另外选择（或可能）；非此即彼；二者择一地；作为一种选择

Rain Barrels

Rainwater harvesting is the ancient practice of capturing and using rainwater. About half of all household water use in the summer goes to watering lawns and gardens, washing cars and even driveways. It makes sense, and saves money to use rainwater instead of tap water whenever possible.

The use of rain barrels, installed at the bottom of a downspout, is the most common form of rainwater harvesting. The rain barrel collects rainwater that flows off your roof, storing it for later use. Rain barrels come in a variety of sizes and colours, and can be easily incorporated into your existing landscape.

What's so Great About Rainwater?

Rainwater Saves Money!
By using rainwater in your yard and garden, you can save money on your municipal water bill.

Rainwater is Better for Plants!
Plants prefer rainwater. Rainwater is warmer and softer than tap water, and it hasn't been treated with chlorine.

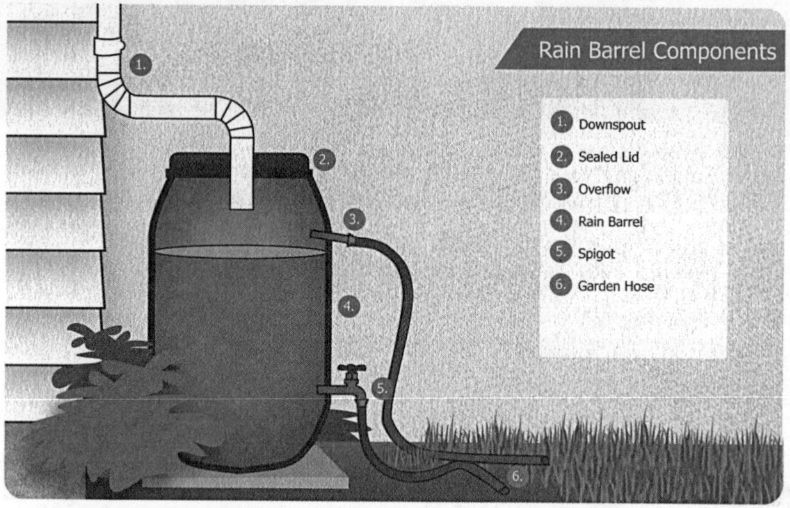

Rainwater is a Resource!
Rain barrels provide a ready supply of water during the hot, dry days of summer, reducing the stress on our waterways and groundwater reserves.

Getting Started:

What's Your Style?
There are many different sizes, shapes and styles of rain barrels. Your regional municipality may sell rain barrels at a subsidized rate at certain times of the year, or you may want to see what your local hardware store or garden center has available.

IS A RAIN BARREL RIGHT FOR ME?

	YES	NO
Do you have a surface from which you can collect rain water? (roof of a home, garage or shed)	☐	☐
Do you have visible eaves troughs around the perimeter of your roof?	☐	☐
Are your downspouts disconnected? If you're unsure, refer to the Downspout Disconnection section on page 24.	☐	☐
Do you have enough space to install a rain barrel on your property (approximately 2 metres around your downspout with no obstructions)?	☐	☐
Do you have space to redirect the overflow from your rain barrel onto your property?	☐	☐

If you answered 'YES' to all of the above, you are ready to install your rain barrel! Read on to learn how.

雨水桶

雨水收集是一种古老的获取并利用雨水的方法。每年夏季，大约一半的家庭用水都消耗于浇灌草坪、花园，清洗车辆甚至车道。尽可能地用雨水代替自来水不仅能节约水量，还能省钱。

在落水管底部安装雨水桶是最常见的雨水收集方式。雨水桶收集从屋顶流下来的雨水并储存起来，供以后使用。雨水桶的大小和颜色各不相同，可以很容易与现有庭院景色搭配。

雨水有什么好处？

利用雨水可以省钱！

在庭院和花园中使用收集的雨水，可以节约市政水费。

雨水对植物更有益！

植物喜欢雨水。雨水比自来水温度高、硬度低，而且没有经过氯处理。

雨水是一种资源！

在炎热干燥的夏季，雨水桶可以提供充足的水量，减轻了市政供水和地下水开采的压力。

开始：

你想要的风格是什么？

雨水桶有各种尺寸、形状和风格。当地政府可能会在一年的某些时候以补贴价格出售雨水桶，或者可以到当地的五金店或花园建造配件中心选购。

你家的庭院适合使用雨水桶吗？

	是	否
1. 你有可用的雨水收集的表面（屋顶、车库顶或棚顶）吗？	□	□
2. 你的屋顶周围屋檐可以集雨吗？	□	□
3. 你的落水管断开了吗？如果不确定，请参考第 25 页“断开的落水管”部分的内容。	□	□
4. 你是否有足够的空间安装雨水桶（落水管周围大约 2 米范围内没有障碍物）？	□	□
5. 你的庭院是否有空间容纳雨水桶中溢流出来的雨水？	□	□

对于上述所有问题，如果你都回答“是”，那么你已经准备好安装雨水桶了！继续往下阅读，学习具体做法。

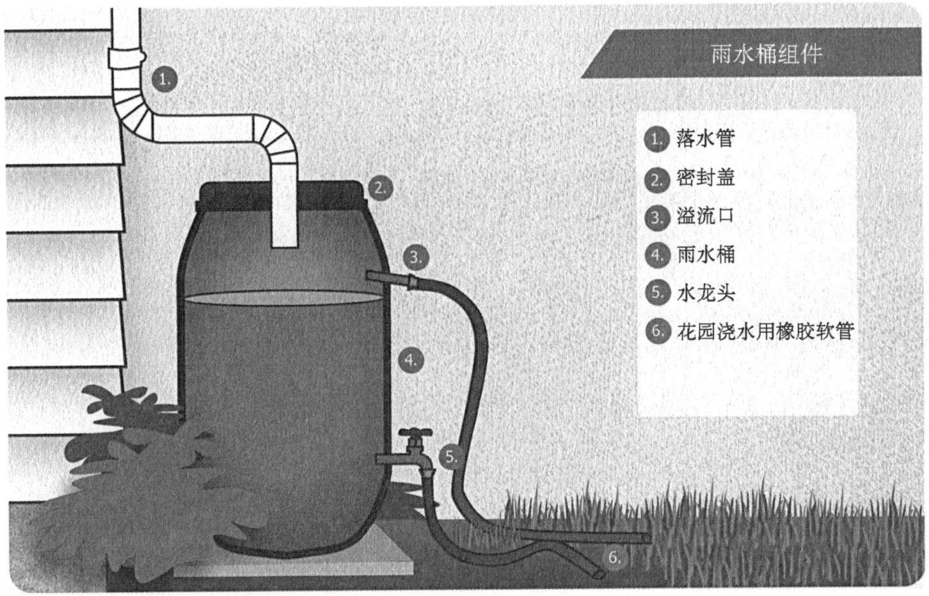

专业词汇

rain barrels 雨水桶
rain water harvesting 雨水集蓄
tap water 自来水
downspout /ˈdaʊnspaʊt/*n.* 落水管（将雨水从屋顶排至水沟）
incorporated into 融入
landscape/ˈlændskeɪp/*n.* 风景；风景画；景色；山水画；乡村风景画；地形；（文件的）横向打印格式
municipal water bill 市政水费
chlorine/ˈklɔːriːn/*n.* ［化学］氯（17 号化学元素）
groundwater reserves 地下水储量
regional municipality 市政当局
hardware store 五金店
property/ˈprɒpəti/*n.* 性质，性能；财产；所有权
approximately /əˈprɒksɪmətli/*adv.* 大约，近似地；近于
obstruction/əbˈstrʌkʃn/*n.* 障碍；阻碍；妨碍

Read about the benefits of each type of rain barrel to determine which is best for you.

Collapsible Rain Barrel – Made from pliable PVC, these rain barrels are easy to transport, store and assemble. They are great in areas where the ground is not level, as they can rest on any terrain.

Photo Credit: Lee Valley Tools Ltd.

Plastic Rain Barrel – Most likely the type you will find at your local hardware store and/or what your regional municipality sells. Made from durable hard plastic, they sit about waist high and hold approximately 200 litres of water.

Riversides 'River Safe Rain Barrel'- This tough, recycled, 500 litre barrel resists fading, warping and cracks. It has over twice the capacity of most other rain barrels on the market and its shape ensures stability while providing height for gravity fed usage. It also has a closed design and mosquito proof filters. (www.riversides.org)

Wooden Barrel – These are usually recycled wine barrels that have to be converted into a rain barrel.

Building Your Own Rain Barrel
It's easy to build your own rain barrel, and it allows you to design it to fit your space. The tools and materials are available at most hardware stores. Barrels may be available at select hardware stores, and at businesses such as material recyclers and food importers.

For step by step instructions on how to build your own rain barrel, visit www.ottawa.ca/residents/water/waterwise/outdoors/lgt/rain_barrel_en.html. For safety, always use personal protective equipment such as gloves and safety glasses.

LOCAL PLACES THAT SELL RAIN BARRELS AROUND THE GTA

- Region of Peel
- Eco Caledon
- Riversides
- Durham Sustain-ability
- Rainbarrel.ca
- Local hardware and landscape supply stores

IMAGINE YOUR LANDSCAPE WITH A RAIN BARREL (OR TWO...)

Referring back to the property sketch you drew at the beginning of the guidebook (page 8), consider what your yard would look like with one or more rain barrels. Where will you install them, where will you be using the water, which type do you like?

了解每种类型雨水桶的优点，以确定哪一种最适合。

可折叠的雨水桶——这种雨水桶由柔韧性强的 PVC 材料制成，便于运输、储存和组装。这种雨水桶在地面不平整的地方甚至是任何地形条件下都可以使用。

图片来源：李山谷用具有限公司

塑料雨水桶——当地的五金店或所在地区的市政当局销售的大多是这种类型的雨水桶。由耐用的硬塑料制成，齐腰高，大约可容纳 200 升水。

超容量雨水桶——这种雨水桶坚硬、可回收、容量达 500 升，还可以抵抗褪色、翘曲和裂纹等问题。它的容量超过市场上大多数雨水桶的两倍以上，形状设计方面能够保持水桶的稳定性。这种雨水桶是封闭式设计的，还设有防蚊过滤器。(www. riversides. org)

木制雨水桶——通常由回收的酒桶改造而成。

自制雨水桶——自制雨水桶很容易，而且自己设计更能满足个性化需求。所需的工具和材料在大多数五金店都能买到。在有些五金商店以及物料回收商和食品进口商那里，可以直接买到水桶。

关于如何自制雨水桶的步骤说明，请访问 www. ottawa. ca/residents/water/waterwise/outdoors/lgt/rain _ barrel _ en. html. 为了安全起见，请使用手套和护目镜等个人防护用品。

GTA 附近卖雨水桶的地方

- 皮尔地区
- 艾柯卡列登
- 河畔
- 杜伦大学可持续发展研究中心
- rainbarrel. ca
- 本地五金店和园林用品店

想象一下拥有一两个雨水桶后庭院的风景

回顾这本手册刚开始部分（第 9 页）你画的那张草图，想象一下庭院里安装一个或多个雨水桶后的样子。你要把雨水桶安装在哪里？要把水用在哪里？喜欢哪种类型的雨水桶？

专业词汇

collapsible/kə'læpsəbl/*adj*. 可折叠的；可拆卸的

pliable/'plaɪəbl/*adj*. 柔韧的；柔软的；圆滑的；易曲折的

terrain/tə'reɪn/*n*. [地理] 地形，地势；领域；地带

durable/'djʊərəbl/*adj*. 耐用的，持久的

waist high 齐腰高

recycled/ˌriː'saɪkld/*adj*. [环境] 回收利用的；可循环再造的

fading/'feɪdɪŋ/*n*. 褪色；衰退；凋谢

warping/'wɔːpɪŋ/*n*. 翘曲

cracks/kræks/*n*. 龟裂（crack 的复数形式）

mosquito/mɒ'skiːtəʊ/*n*. 蚊子

material recycler 物料回收商

food importers 食品进口商

personal protective equipment 个人防护装备

sketch/sketʃ/*n*. 素描；略图；梗概

Location

Looking at the diagram of a typical property below, circle the rain barrels that you think have been placed in a suitable area. Draw an "X" through those that are not in a suitable area. For answers, refer to page 36.

位置

看看下面的典型房屋结构图，把认为适合放置雨水桶的位置圈起来，在认为不适合的地方画一个“×”。有关位置的确定请参阅第 37 页。

专业词汇

diagram/ˈdaɪəgræm/*n.* 图表；图解

YES Locations

Next to a garden – By placing your rain barrel under a downspout located close to your garden, you will be sure to always have a ready supply of water close to where you need it. **TIP**: Try hooking up a soaker hose to your rain barrel for slow drip irrigation.

Next to your patio or deck – By placing your rain barrel under a downspout located next to your patio or deck, you can easily water hanging baskets or planters.

Under garden shed downspout – Why not make use of the water that runs off your garden shed or garage – if there are downspouts located here you can easily install a rain barrel next to these structures.

On a slope away from your house – In order to prevent any pooling of water, it is advised to place your barrel in an area where the grade is sloped away from the foundation of the house.

NO Locations

On your driveway, pathway, front porch or other hard surface – Rain barrels should not be installed on large paved surfaces where there is no appropriate surface to direct the overflow.

On a slope toward your house – If at all possible, this situation should be avoided. When the grade of the land is sloping towards the foundation of your home, the chance of water entering your basement increases.

MAYBE Locations

Next to the pathway – If installing a rain barrel next to a walkway or path, you must ensure there is room to direct the overflow to an appropriate permeable surface, such as a garden or lawn.

In between 2 houses – If installing your rain barrel between your house and your neighbour's, ensure there is enough space to move around the barrel, and ensure that the overflow is directed an appropriate distance downhill from both homes (2 metres).

WHAT TO DO WITH THE OVERFLOW

All rain barrels come with an overflow system. This system ensures that once your rain barrel is full, any additional water will be directed to a suitable location away from your home, rather than overflowing over the side of the rain barrel and potentially causing flooding or other water issues. On most rain barrels, the overflow system is a pipe or hose attached to a small opening near the top of the barrel. The pipe or hose should be situated so that water running through it is directed downhill and away from the foundation of the home. Rain barrel overflow can also be directed towards your lawn, planted areas, or even a rain garden (refer to the Rain Gardens chapter for more information).

适合的位置

花园旁边——把雨水桶放在靠近花园的落水管处，需要用水的时候随时有水可用。**小贴士**：试着在你的雨水桶上连接一根软管，这样就可以进行缓慢的滴灌。

天井或露台旁边——把雨水桶放在天井或露台旁边的落水管处，方便浇灌吊篮或花盆中的植物。

花园棚屋落水管处——如果这里设置有落水管，为什么不把花园棚屋或车库顶流下来的雨水利用起来呢？你可以很容易地在这些建筑物旁边安装雨水桶。

斜离房屋的斜坡——为了防止积水，建议把雨水桶放在斜离房屋地基的区域。

不适合的位置

车道、通道、前门廊或其他坚硬的地面——雨水桶不应安装在这些宽阔的铺砌路面上，因为这种路面不适合引导雨水径流。

斜向房屋的斜坡——如果可能的话，这种情况应该避免。当斜坡向房屋地基倾斜时，地下室被淹的可能性较高。

需要考虑的位置

通道旁边——如果在人行道或小路旁边安装雨水桶，必须确保有足够的空间将雨水径流引导到合适的可渗透地面，如花园或草坪。

两个房子之间——如果在你和邻居的房子之间安装雨水桶，你必须确保雨水桶的周围有足够的空间，并确保雨水径流与两栋房子之间都有一定的距离（不小于 2 米）。

如何处理溢流？

所有的雨水桶都设有溢流系统。一旦雨水桶被装满，多出来的雨水就将被导流到远离房屋的合适位置，而不是溢出在雨水桶周围引发水淹或其他问题。大多数雨水桶的溢流系统是通过连接在桶顶附近开口上的一根水管或橡胶软管实现的。水管或橡胶软管安装在这个位置，多余的雨水可以被导流到远离房屋的地方，如草坪、种植区域，甚至是雨水花园（更多信息请参考“雨水花园”章节）。

专业词汇

hanging basket 挂篮；吊篮
planter/ˈplɑːntə(r)/*n.* 花盆
foundation of the house 地基
basement/ˈbeɪsmənt/*n.* 地下室；地窖

How to:
Install Your Rain Barrel

Tools Required:
- Shovel
- Level
- Sand
- Bricks or patio stones
- Raised platform (optional)
- Tin snips or hacksaw
- Strapping (optional)
- Downspout elbow
- Hose

IMPORTANT

Refer to the section on 'Downspout Disconnection' to make sure you have completed all necessary steps to disconnect your downspout prior to installing your rain barrel.

Steps

1 Using a shovel, level the ground where the rain barrel will sit.

2 On the leveled area, spread a one inch thick layer of ground limestone or sand, then proceed to lay bricks or patio stones on top to create a strong base.

2b (Optional)
Install a raised platform under your rain barrel. This will elevate the barrel, allowing room to place a watering can underneath the spigot. It will also assist gravity in creating water pressure if you are planning to attach a hose.

3 Place the rainbarrel against the downspout so you can measure its height from the ground. Mark the downspout six inches above the top of the rain barrel.

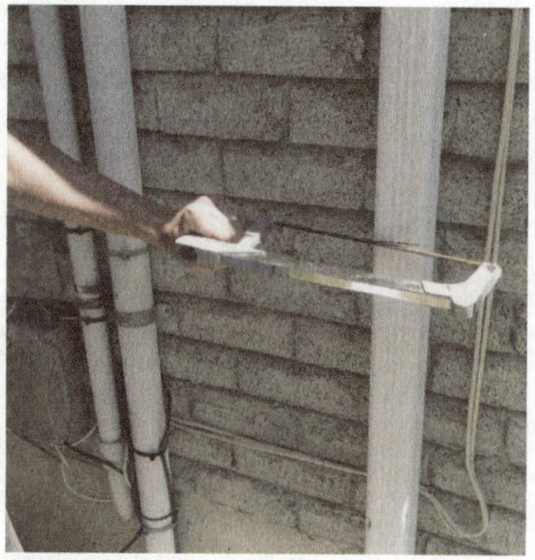

4 Cut the downspout with metal sheers or a hacksaw. Keep the cut piece of downspout. You will want to use it when over wintering your rain barrel. Refer to "Maintenance" on page 42 for more information.

5 Attach an 'elbow' onto the newly cut downspout and screw it in with two self threading screws.

如何安装雨水桶?

所需工具:

- 铲子
- 水平尺
- 砂子
- 砖或铺路石
- 垫高平台（可选）
- 金属剪或钢锯
- 胶带（可选）
- 肘型弯管
- 橡胶软管

步骤

1. 用铲子平整地面。

2. 在平整好的地面上铺 1 英尺（约等于 30.5 厘米）厚的石灰石或砂子，然后在上面铺上砖或铺路石以形成坚固的地基。

2b. （可选）在雨水桶下面安置一个凸出的平台。这样可以抬高桶的位置，使水龙头下有足够的空间可以放置浇水罐。如果计划安装软管，这样也可以帮助增加水压。

3. 将雨水桶放置在落水管正下方，测量自地面到桶顶的高度。在雨水桶上方六英寸（约等于 15.24 厘米）处标注落水口的位置。

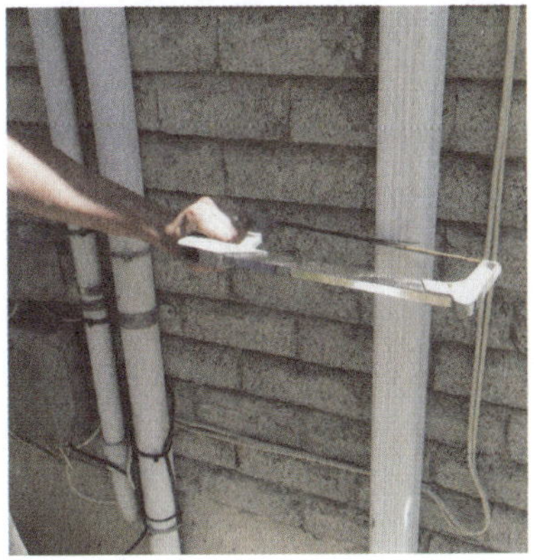

4. 用金属剪或钢锯将落水管剪开。保留剪下的部分，以备越冬的时候用到它。更多信息请参阅第 43 页有关“维护”的内容。

5. 将“肘型弯管”安在剪切后的落水管上，并用两个自攻螺丝拧紧。

重要提示

请参阅“断开的落水管”部分的内容，以确保在安装雨水桶之前你已经完成断开落水管的所有必要步骤。

专业词汇

limestone/ˈlaɪmstəʊn/n. [岩] 石灰岩
watering can 洒水壶；喷壶
water pressure 水压
self threading screws 自攻丝螺钉
disconnect/ˌdɪskəˈnekt/vt. 拆开，使分离

6 Place the rain barrel on the platform.

7 Ensure the newly installed elbow from the downspout is directed into the top opening of the rain barrel so the water flows into it.

8 Connect the overflow valve to a hose. Direct it away from the house towards a planted area or another rain water landscape project in this guide. (Refer to the "Connecting the Drops" chapter, page 96)

9 Make sure the spigot at the bottom of the rain barrel is closed so that the water does not run directly out of the barrel.

10 Your rain barrel is now ready for the next rainfall.

ATTACHING A SOAKER HOSE

A soaker hose is basically a hose with holes in it. By attaching one to the spigot on your rain barrel, you can water your lawn or garden with no effort at all. The water from the rain barrel will slowly move through the hose, through the holes and into the soil.

6. 把雨水桶放在垫高平台上。

7. 确保新安装的肘型弯头进入雨水桶顶部开口处，这样雨水才会流入桶内。

8. 将溢流阀连接到软管上。将水流引流到远离房屋的种植区域或本手册中的其他雨水景观项目区（请参阅第 97 页“雨水的连通路径”章节）。

9. 确保雨水桶底部的水龙头关闭，这样水就不会直接从桶中流出。

10. 你的雨水桶可以开始集蓄雨水了。

关于渗水软管

渗水软管是一根有洞的软管。在雨水桶的水龙头上安装一根渗水软管，就可以毫不费力地浇灌草坪或花园。雨水桶里流出的水会慢慢通过渗水软管上的洞渗入土壤。

专业词汇

valve/vælv/*n.* 阀门

soaker hose 渗水软管

Maintaining your rain barrel

A rain barrel can be a great addition to any home. But like many outdoor products, it needs to be maintained. Although the maintenance of a rain barrel is not complicated, it is important in both extending the life of your rain barrel and keeping the water that it collects clean.

Regular Maintenance

Your rain barrel should be inspected regularly, especially after a large rain event. Be sure to check for tears in the bug screen, cracks or leaks in the spigot, hose, valves or barrel itself. Leaf and twig debris should be cleaned away from the top of the barrel to prevent clogging. Hoses should also be checked for obstructions.

Cleaning

Your rain barrel should be cleaned once before use, and then once a year after that. Vinegar or another non-toxic cleaner should be used on both the inside of the barrel and the spigot. For best results, scrub the barrel with a non-abrasive brush. Be sure to rinse the barrel thoroughly after cleaning and dispose of the water onto your lawn and not down the storm drain.

How to:

Overwintering

Rain barrels should be disconnected from your downspouts in the fall, dried and properly stored. If a rain barrel is left out over the winter with water in it, the water will freeze and cause cracking and other damage to the barrel.

When overwintering your rain barrel, there are a few steps you will need to take.

1. Disconnect the downspout from the rain barrel.

2. Using the piece of downspout that you saved when installing your rain barrel (refer to step 4 of "Install your Rain Barrel"), return it to its original configuration. You will need to use a connector. Alternatively, you can connect a flexible downspout diverter. Be sure the downspout is angled away from the foundation of your house, at least 2–3 metres.

3. Remove the hose and mesh screen from your barrel. Ensure all water is drained and store them in a dry place.

4. Drain the rain barrel.

5. Turn the barrel upside down to prevent water from collecting in it. For best results, store it in an enclosed area like a garage or shed.

Things to Consider

Safety

Full rain barrels can weigh upwards of 500 pounds, so be sure your barrel is sturdy and not prone to tipping when in place. To prevent tipping, you can use a water heater strapping kit to secure it to a nearby wall or fence post.

Mosquitoes

West Nile Virus is mainly transmitted to people through the bite of an infected mosquito. While it is very unlikely that you will be exposed to an infected mosquito, it is necessary to take certain precautions with your rain barrel to ensure that you aren't creating a breeding ground. Mosquitoes require water that is left undisturbed for a minimum of 7 days in order to successfully breed. Therefore it is a good idea to use the water regularly.

The best solution is to stop any mosquitoes from entering your rain barrel in the first place. All rain barrels should come with a bug screen or netting on the top. If you are building your own rain barrel, ensure that you don't skip this vital step. Fiberglass window screening works well and comes in convenient rolls. Be sure to check the screen periodically to ensure it is in good repair.

雨水桶的维护

雨水桶是每个家庭的必备品。但像许多户外用品一样，它也需要维护。虽然雨水桶的维护并不复杂，但对延长雨水桶的寿命和保持桶内雨水的清洁都很重要。

定期维护

应做到定期检查雨水桶，特别是在大雨之后。一定要检查防蚊过滤器内有无杂物，水龙头、软管、阀门或桶身有无裂缝或渗漏。桶顶的树叶和树枝碎屑应及时清除以防止堵塞。软管也应该检查是否堵塞。

清洁

雨水桶应在使用前清洗一次，之后每年清洗一次。桶的内部和水龙头都应使用醋或其他无毒清洁剂清洗。为了达到最佳效果，可以用软毛刷来擦洗。擦洗后一定要把桶彻底冲洗干净，冲洗的水可以倒在自家草坪上，而不能倒入下水道中。

怎么做？

如何过冬

秋天，雨水桶应从落水管处拆卸下来，干燥并妥善保存。如果雨水桶在冬天被放置室外，桶中的水会结冰，导致桶身破裂或产生其他损坏。

越冬时，你需要按照以下步骤处理：

1. 把落水管从雨水桶中拔出来。

2. 将安装雨水桶时保留下来的那截落水管（参考雨水桶安装第 4 步）恢复到原来的位置。你可以使用接头连接一个灵活的防溅垫，确保落水口向远离房屋地基的方向倾斜，至少2～3米。

3. 从雨水桶上拆下软管和筛网，确保所有的水都被排干，存放在干燥的地方。

4. 排干雨水桶。

5. 把桶倒过来，防止水在桶底积存。为了达到最好的保存效果，可以把它存放在一个封闭的空间，比如车库或棚屋。

需要考虑的事情

安全

装满水的雨水桶重量可能超过 500 磅（约等于 227 千克），所以要确保雨水桶的稳固性，不能倾斜。为防止意外，可以使用捆扎工具把桶固定在附近的墙壁或栅栏上。

蚊子

西尼罗河病毒主要通过受感染蚊子的叮咬传播给人们。虽然被受感染的蚊子叮咬的可能性不大，但有必要采取一定的预防措施，确保不会人为创造蚊子繁殖的场所。蚊子需要至少 7 天不受干扰的水环境才能成功繁殖，所以，经常使用桶中的水可以有效避免蚊子繁殖。

最好的解决办法是：首先阻止任何蚊子进入雨水桶中，所有雨水桶顶部都应有一个防蚊过滤器或网盖。如果你正在做雨水桶，请确保不会跳过这一重要步骤。玻璃纤维窗纱性能良好，并设有方便的辊子。一定要定期检查窗纱，确保它处于良好的使用状态。

专业词汇

maintenance/'meɪntənəns/*n.* 维护，维修；保持；生活费用
twig/twɪg/*n.* 小枝；嫩枝；末梢
clogging/klɒgɪŋ/*n.* 堵塞
vinegar/'vɪnɪgə/*n.* 醋
non-toxic cleaner 无毒清洗剂
storm drain 雨水沟；雨水道
configuration/kənfɪgə'reɪʃn/*n.* 配置；结构；外形
connector/kə'nektə(r)/*n.* 连接器，连接头
alternatively/ɔːl'tɜːnətɪvli/*adv.* 非此即彼；二者择一地；作为一种选择
flexible/'fleksəbl/*adj.* 灵活的，柔韧的，易弯曲的
mesh screen 筛孔，网筛
prone to 有…倾向的
transmit/trænz'mɪt/*vt.* 传输；传播；发射；传达；遗传
exposed to 接触，暴露于
precaution/prɪ'kɔːʃn/*n.* 预防，警惕；预防措施
undisturbed/ˌʌndɪ'stɜːbd/*adj.* 安静的；镇定的；未被扰乱的；泰然自若的
a minimum of 至少；最少的
fibreglass/'faɪbəglɑːs/*n.* 玻璃纤维；玻璃丝
window screening 窗纱
periodically/ˌpɪərɪ'ɒdɪkli/*adv.* 定期地；周期性地；偶尔；间歇

West Nile Virus
To find more information about West Nile Virus, contact your local Regional Public Health Division.

York Region: 1-800-361-5653 to speak to a Public Health Inspector

Peel Region: 1-905-799-7700 to speak to Peel Public Health, or visit www.peel-bugbite.ca

Toronto: 1-416-338-7600 to speak to Toronto Public Health, or visit www.toronto.ca/health/westnile

Durham Region:
1-888-777-9613 ext.2188

Flooding / Water Pooling
The overflow hose should ALWAYS be directed at least 3 metres away from the house to prevent water from pooling near the foundation. Completing this step will decrease the chance of water reaching your foundation.

Beyond the Basics

Art for Your Yard

Painted rain barrels are becoming popular, allowing you to add an artistic element to your garden. Get creative!

How to:

1. Wipe down the barrel with a one-to-one mixture of vinegar and water.
2. Rough the surface of the barrel with a piece of fine to medium grit sand paper.
3. Apply a coat of latex bonding primer.
4. Paint your design! Exterior latex paint is preferred, but you can also use any leftover cans of paint you have around the house, or basic artist's paints.
5. Seal and protect your barrel with a coat of polyurethane.
6. Install your new "work of art for the garden" and enjoy!

WHAT TO DO WITH ALL THAT RAIN WATER...

How Much Water Are We Talking About Here?
The amount of rainwater that runs off a roof in a year is enough to flush a 6-litre toilet 12,000 times - that's more than 32 flushes per day!

Now that you have collected all this rain water, it's time to use it! The simplest way to use the rainwater in your barrel is to fill a watering can and water any gardens, potted plants or hanging baskets on your property. Another option is to attach a soaker hose, which can be used to slowly water a nearby garden. A typical rain barrel can often fill up after just one rain storm, so be sure to use the water collected in your barrel on a regular basis.

IMPORTANT

Keep in mind that the water collected inside the rain barrel is non-potable water and should not be consumed. However, it is a safe, natural source of water for your grass and flower gardens.

西尼罗河病毒

想了解更多有关西尼罗河病毒的信息，请联系当地的公共卫生部门。

约克地区：1-800-361-5653 咨询公共卫生检查员。

皮尔地区：1-905-799-7700 咨询公共卫生部门或访问 www.peel-bugbite.ca。

多伦多：1-416-338-7600 咨询多伦多公共卫生部门或访问 www.toronto.ca/health/westnile。

达勒姆地区：1-888-777-9613 ext.2188

洪水/积水

溢流软管出水口应始终位于离房屋至少 3 米远的地方，以防止地基附近积水。这一步骤对于防止房屋地基积水至关重要。

拓展知识

庭院艺术

彩绘雨水桶变得越来越流行，可以给花园增添一些艺术元素，这很有创意！

怎么做呢？

1. 醋和水按 1∶1 的比例混合，将桶身擦干净。
2. 用中等粒度的砂纸将桶身磨毛。
3. 涂一层乳胶漆。
4. 开始彩绘设计！外墙乳胶漆是首选，但也可以使用家里剩下的任何其他油漆或是普通的绘画颜料。
5. 涂一层聚氨酯漆以保护雨水桶。
6. 安装“花园艺术作品”吧！

如何处理这些雨水……

这里有多少水？

屋顶一年流下的雨水总量足以让一个 6 公升的马桶冲水 12000 次——每天超过 32 次！

既然已经收集了所有的雨水，是时候使用了！使用雨水桶里的水最简单的方法就是：装满洒水壶，浇灌花园、盆栽植物、或者吊篮植物。另一种选择是：安装一根渗水软管，慢慢地给附近的花园浇水。一个典型的雨水桶通常在一场暴雨之后就会被灌满，所以一定要经常使用桶里的水。

重要提示

请记住，雨水桶里收集的水是不能饮用的，但对于草坪和花园植物来说却是一种安全的、天然的水源。

专业词汇

Public Health Division 公共卫生部门
sand paper 砂纸
latex bonding primer 乳胶粘合底漆
polyurethane/ˌpɒlɪˈjʊərəθeɪn/*n.* 聚氨酯；聚亚安酯
on a regular basis 定期地，经常地

Rain Gardens

Rain gardens are shallow, sunken gardens, designed to collect rain water that runs off your roof, driveway or patio. Rain gardens can be planted with any combination of shrubs, grasses and flowers, and look just like a regular garden. The most important part of a rain garden is what you can't see – the loose, deep soil underneath that absorbs and filters rain water. Rain gardens are easy to build and maintain, even if you don't have a green thumb!

Benefits of a Rain Garden

- Absorbs excess water on your property during rain storms.
- Filters oil, fertilizer and other pollutants out of stormwater before they reach nearby storm drains and are carried to local rivers and lakes.
- Increases the amount of water that soaks into the ground to recharge the local groundwater supply.
- Provides habitat for wildlife like birds, butterflies and beneficial insects.
- Is a beautiful, unique garden that will make you the envy of the neighbourhood.

Credit: The City of Calgary

Getting Started

The best gardens start with a well laid out plan. This is especially true for rain gardens. There are three things to consider when planning your new rain garden:

- *Location*
- *Style*
- *Size*

Location

Where will the water come from? Water can be carried to your rain garden in a variety of different ways. It can be carried through an extension on the end of your downspout, through an infiltration trench lined with plants or decorative rocks (refer to page 64), or through the overflow pipe from your rain barrel (refer to page 36).

In section 2, you assessed your property to locate your downspouts. Redirecting a downspout is the easiest way to carry water to your rain garden. Depending on the size and location of your property, you can even consider directing more than one downspout towards your rain garden.

IS A RAIN GARDEN RIGHT FOR ME?

	YES	NO
Do you have a low lying area on your property where water naturally flows during a heavy storm?	☐	☐
Are you interested in installing a new garden, or do you have an existing garden that you would like to redesign?	☐	☐
Is your potential garden space at least 3 metres away from any building foundation on your property?	☐	☐
Is your potential garden space relatively flat? *(between 1% and 5% slope – see page 12 for slope calculations)*	☐	☐
Is there a source of water to feed your potential garden space, such as a downspout or a rain barrel?	☐	☐

If you answered 'YES' to all of the above, you are ready to install your rain garden! Read on to learn how.

If you didn't answer 'YES' to all of the above, there may be a different garden project that would work for your property. Turn to page 64 to learn about soakaways.

雨水花园

雨水花园是一种很浅的下沉式花园，可以用来收集从屋顶、车道或露台上流下来的雨水。雨水花园中可以种植任何灌木、草和花，看起来就像一个普通的花园。雨水花园最重要的部分是看不到的——吸收和过滤雨水的松散的深层土壤。即使没有园艺技能，雨水花园也很容易建造和维护。

雨水花园的好处

- 在暴雨期间吸纳庭院多余的雨水。
- 在雨水流入附近的排水沟进入当地的河流和湖泊之前，将石油、肥料和其他污染物从雨水中过滤出来。
- 增加渗入地下的水量，以补充当地的地下含水层。
- 为鸟类、蝴蝶和益虫等野生动物提供栖息地。
- 它是一个美丽的、独特的花园，会让你成为邻居们羡慕的对象。

开始

最好的花园是从精心设计开始的，对于雨水花园来说更是如此。在规划新的雨水花园时，有三件事需要考虑：位置、风格、大小。

位置

水从哪里来？ 水可以以各种不同的方式进入雨水花园。它可以通过落水管末端的延伸装置，经过一个有植物或装饰性岩石的渗透沟（参考第 65 页）或者通过雨水桶的溢流管道进入雨水花园（参考第 37 页）。

在第 2 部分“场地评估”中，你对家里安装落水管的位置进行了评估，而利用落水管把水引到雨水花园是最简单的方法。根据房屋的

图片来源：卡尔加里市

大小和位置，可以考虑设置多个落水管往雨水花园引水。

我家适合建雨水花园吗？

	是	否
• 你家是否有一个低洼的区域，在下雨时雨水会自然地流进去？	☐	☐
• 你有兴趣建造一个新花园吗？或者重新设计家里的花园吗？	☐	☐
• 花园的位置是否离房屋地基至少 3 米远？	☐	☐
• 你的花园相对平坦吗？（坡度在 1%到 5%之间——参阅第 13 页坡度计算）。	☐	☐
• 你是否有灌溉花园的水源，比如落水管或雨水桶？	☐	☐

对于上述所有问题，如果都回答“是”，那么你已经准备好建造雨水花园了！继续往下读，学习如何建造。

对于以上所有问题，如果不全回答“是”，那么你可以考虑建造其他花园项目。请翻到第 65 页了解渗水坑。

专业词汇

shallow/ˈʃæləʊ/*adj*. 浅的

sunken/ˈsʌŋkən/*adj*. 沉没的；凹陷的；比周围低的

combination/ˌkɒmbɪˈneɪʃn/*n*. 结合；组合；联合

shrub/ʃrʌb/*n*. 灌木；灌木丛

loose/luːs/*adj*. 宽松的；散漫的；不牢固的；不精确的

green thumb 园艺技能

fertilizer/ˈfɜːtɪlaɪzə/*n*. 肥料

habitat/ˈhæbɪtæt/*n*. [生态] 栖息地；（动植物的）[经] 产地（habitat 的复数形式）

infiltration trench lined 渗滤沟排

decorative/ˈdek(ə)rətɪv/*adj*. 装饰性的；装潢用的

slope/sləʊp/*n*. 斜坡；倾斜；斜率

soakaway/ˈsəʊkəweɪ/*n*. 渗水坑

Rain gardens can be placed in more than one location, depending on your space and how ambitious you're feeling. For example, you can direct water from your rain barrel to a rain garden in your backyard, and water from your downspouts to a rain garden in your front yard.

There are a few locations on your property that ARE NOT appropriate for rain gardens:

- To prevent water from getting into foundations and causing basement flooding, rain gardens should always be at least three metres away from your home, garage and any neighbouring properties.
- Rain gardens should never be placed within 4 metres of a septic bed, and if uphill of a septic system, they should be at least 15 metres away.
- Rain gardens placed near any type of steep slope (more than 15% - refer to page 12 for instructions on calculating slope) have the potential to cause land slides because of the excess water being absorbed into the soil. Rain gardens should not be installed within 15 metres of any steep slopes on or around your property.
- Before doing any digging for this project, you must locate and mark all underground utilities. Some utilities are quite shallow and you can be held liable if you hit them without obtaining the proper dig permits first.
- Areas of poor drainage are not likely good options for rain gardens. However, you can potentially amend the soil to make it suitable for a rain garden. Refer to page 14 for more information on soil and infiltration rates.

What's your style?

There are many different styles of rain gardens to choose from. Whether you like a formal landscape, or a more casual country look, you can create a rain garden that you will love. The options are endless!

Photo Credit: Sean James, Fern Ridge Landscaping

LOCATION CHOICES

Refer back to the property sketch you did back on page 8 and sketch in one or two areas where you might like to see a rain garden.

IMPORTANT

CALL BEFORE YOU DIG!

ON1CALL
CALL BEFORE YOU DIG
It's The Law!
www.on1call.com
1-800-400-2255

TO LEARN MORE:

There are a few demonstration rain gardens in the Greater Toronto Area that you can visit for inspiration:

- Humber Arboretum
  (205 Humber College Boulevard, Toronto)
- County Court demonstration rain garden
  (71 Turtlecreek Boulevard, Brampton)
- Kortright Centre for Conservation, Archetype House Gardens
  (9520 Pine Valley Drive, Vaughan)
- Markham Museum
  (9350 Markham Road, Markham)
- Lake Wilcox demonstration rain gardens
  (20 Wheatsheaf Street, Richmond Hill and 95 Wheelright Drive, Richmond Hill)

雨水花园可以设置在不同的地方，这取决于庭院的空间和你的审美。例如，你可以把雨水从雨水桶直接引到后院的雨水花园，也可以把落水管流下的雨水引到前院的雨水花园。

有一些位置不适合建造雨水花园：

- 为了防止雨水进入地基淹没地下室，雨水花园应该离房屋、车库和任何邻近的建筑至少3米远。
- 雨水花园不应设置在距化粪池4米范围内，如果位于化粪池系统的上游坡位置，则应至少相距15米。
- 因为过量的水被吸收到土壤中，任何陡坡附近的雨水花园（超过15%——参考第13页的坡度计算相关介绍）都有可能导致滑坡。所以，雨水花园不应设置在你家或周围任何陡峭山坡的15米范围内。
- 在进行挖掘作业之前，必须找到并标记所有的地下设施。有些公用设施埋置得很浅，如果不事先获得相关的挖掘许可，可能要承担相应责任。
- 排水系统差的地区不太适合建造雨水花园。但是可以进行土壤改良处理，使它满足雨水花园设计需要。有关土壤和渗透率的更多信息，请参阅第15页。

你的风格是什么？

有许多不同风格的雨水花园可供选择。无论是喜欢传统的风格，还是休闲的乡村风格，都可以按照自己的喜好设计雨水花园。选项是多种多样的！

弗恩山脊景观　　图片来源：肖恩·詹姆斯

位置选择

回顾一下第9页所绘制的草图，并选取一两个你想建造雨水花园的地方。

重要提示

开挖前先打个电话！

开挖前先打个电话！
这是法律！
www.on1call.com
1－800－400－2255

了解更多知识：

在多伦多地区有一些雨水花园示范基地，你可以去那里寻找灵感：

- 亨伯植物园（多伦多亨伯学院大道205号）
- 州法院雨水花园示范（宾顿市Turtecreek大道71号）
- 科尔特里特保护中心，原型房屋花园（沃恩松谷路9520号）
- 马卡姆博物馆（马卡姆马卡姆路9350号）
- 威尔科克斯湖示范雨水花园（里士满街20号，里士满山右转道95号）

专业词汇

uphill/ʌpˈhɪl/*adj*. 上坡的，向上的
septic system 化粪池系统
land slide 山崩，滑坡，塌方
demonstration/ˌdemənˈstreɪʃn/*n*. 示范
inspiration/ˌɪnspəˈreɪʃn/*n*. 灵感；鼓舞；吸气；妙计

WORKSHEET - Sizing Your Rain Garden

In order to determine the best size for your rain garden, you need to know how much water will be draining to it. The first step is to calculate the total drainage area (your roof area). To do this, you will have to measure the total width and length of your house, and then do the same for the garage (if applicable).

Example:

House: 8 metres x 17 metres = 136 m^2

\+

Garage: 4 metres x 6 metres = 24 m^2

Total roof area of house and garage:

= 160 m^2

Length of House: (A)______________ m

Width of House: (B)______________ m

Total Area of House (multiple A x B): (C) ______________ m^2

Length of Garage: (D)______________ m

Width of Garage: (E)______________ m

Total Area of Garage (multiply D x E): (F) ______________ m^2

Total Roof Area (add together C and F): (G) ______________ m^2

Next, to determine how much water, on average, drains to each downspout, divide the total roof area (G) by the number of downspouts on your house and garage. ($G \div H = J$)

Total number of downspouts: (H) ______________

Total area draining to each downspout: (J) ______________ m^2

Finally, to determine the best size for your rain garden, divide this number by 10. ($J \div 10 = K$)

Your ideal rain garden size: (K) ______________ m^2

This is the recommended size for your rain garden. If you plan to make a larger garden, that's fine, but if you plan to make a smaller garden, you should dig down 10 – 20 centimetres deeper *(refer to page 54 for installation instructions)*.

工作表——确定雨水花园的尺寸

为了确定雨水花园的最佳规模，你需要知道它能消纳多少水。第一步是计算总汇水面积（屋顶面积）。要做到这一点，必须测量房屋的宽度和长度，测量车库的宽度和长度（如果适宜收集雨水的话）。

例如：

房屋：8m×17m=136m²

+

车库：4m×6m=24m²

房屋和车库的总屋顶面积：=160m²

房屋的长度：（A）______________ m

房屋的宽度：（B）______________ m

房屋屋顶面积（A×B）：（C）________ m²

车库的长度：（D）______________ m

车库的宽度：（E）______________ m

车库屋顶面积（D×E）：（F）________ m²

总屋顶面积（C+F）：（G）__________ m²

其次，要确定每根落水管的平均汇水面积，用屋顶总面积（G）除以房屋和车库落水管的总数。（G÷H=J）

落水管总数：（H）__________________

每个落水管的平均汇水面积：（J）______ m²

最后，将这个数字除以 10，得出雨水花园的最佳尺寸。（J÷10）=K

理想的雨水花园尺寸：（K）__________ m²

这是雨水花园的推荐尺寸。如果你打算建一个更大的花园也没问题，但是如果你打算建一个小的雨水花园，应该再挖深 10～20 厘米（参考第 55 页的建造说明）。

专业词汇

recommend/rekə'mend/*vt.* 推荐，介绍；劝告；使受欢迎；托付

A selection of native plants for your rain garden

Rain garden plants need to tolerate both wet and dry conditions. This is because rain gardens experience alternate flooding (following a large rain storm) and drought (during periods of low rainfall). Any combination of flowers, shrubs, grasses or ferns that meet these criteria will do well in a rain garden. Native plants are always recommended for rain gardens because they are well suited to local growing conditions, and support local wildlife including birds and butterflies.

The following is a list of native wild plants for rain gardens, identified based on the light conditions they grow well in:

Sun Partial Sun Shade

Grasses

- big bluestem (*Andropogon gerardii*) — Sun, Partial Sun
- Canada wild rye (*Elymus Canadensis*) — Sun, Partial Sun, Shade
- tufted hairgrass (*Deschampsia cespitosa*) — Sun, Partial Sun
- little bluestem (*Schizachyrium scoparium*) — Sun

Flowering herbaceous plants

- wild columbine (*Aquilegia canadensis*) — Sun, Partial Sun
- butterfly weed (*Asclepias tuberose*) — Sun
- white turtlehead (*Chelone glabra*) — Sun, Partial Sun, Shade
- showy tick-trefoil (*Desmodium canadense*) — Sun, Partial Sun
- spotted Joe-pye weed (*Eupatorium maculatum*) — Sun, Partial Sun
- oxeye sunflower (*Heliopsis helianthoides*) — Sun
- wild bergamot (*Monarda fistulosa*) — Sun, Partial Sun
- black-eyed Susan (*Rudbeckia hirta*) — Sun, Partial Sun
- New England aster (*Symphyotrichum novae-angliae*) — Sun, Partial Sun
- swamp milkweed (*Asclepias incarnata*) — Sun, Partial Sun

Shrubs

- red-osier dogwood (*Cornus sericea*) — Sun, Partial Sun
- eastern ninebark (*Physocarpus opulifolius*) — Sun, Partial Sun
- nannyberry (*Viburnum lentago*) — Sun
- common elderberry (*Sambucus canadensis*) — Sun, Partial Sun

** For a more complete list, refer to the Low Impact Development Stormwater Management Planning and Design Guide, 2010 (see resources guide at the back of this manual for information).*

New England aster

Little bluestem

Joe-pye weed

Butterfly weed

WORKSHEET

Designing Your Rain Garden

There are a few things you might want to think about when designing the look of your rain garden:

- Choose a variety of plants, including shrubs, flowers and grasses, to create variety in colour, height and texture.
- Consider the year-round look of your rain garden – clumping grasses will hold their shape throughout the winter, and many types of shrubs develop striking red branches in the colder months.
- Consider your home's existing landscape, and the landscaping of the surrounding neighbourhood.
- If the rain garden is near the road, consider sight lines and set backs.

Designing Your Rain Garden
continued on next page >

雨水花园中可选择的本土植物

雨水花园植物需要既耐涝又耐旱。这是因为雨水花园需要经历洪水（强降雨后）和干旱（少雨时期）的交替环境。雨水花园中的任何花卉、灌木、草本或蕨类植物，在遇到干湿交替的环境时都应生长良好。本土植物更适宜于雨水花园，因为它们非常适合当地的生长环境，并且与当地的野生动物和谐共生，包括鸟类和蝴蝶。

以下是一份适宜于雨水花园的本土野生植物清单，根据适宜它们生长的光照条件分为：☀喜光照植物、◐半光照植物、●喜阴植物

草本植物

- 大须芒草（桔草） ☀ ◐
- 加拿大野生黑麦（大黄花） ☀ ◐ ●
- 丛生毛草（细叶稷） ☀ ◐
- 小须芒草（裂稃草） ☀

开花的草本植物

- 野生耧斗菜（加拿大耧斗菜） ☀ ◐
- 蝴蝶草（马利筋属晚香玉） ☀
- 白草本植物（草珊瑚） ☀ ◐ ●
- 金钱草（北美舞草） ☀ ◐
- 斑茎泽兰 ☀ ◐
- 牛眼菊向日葵（日光菊） ☀
- 管香蜂草（拟美国薄荷） ☀ ◐
- 黑眼苏珊花（黑心金光菊） ☀ ◐
- 美国紫菀（卷舌菊） ☀ ◐
- 沼泽马利筋（沼泽乳草） ☀ ◐

灌木

- 山茱萸（红瑞木） ☀ ◐
- 东九层皮（金叶风箱果） ☀ ◐
- 糯米树（荚莲属植物） ☀
- 接骨木（美洲接骨木） ☀ ◐

* 要获得更完整的列表，请参阅《雨洪低影响开发管理规划和设计指南》（2010 年）（有关信息，请参阅本手册后面的参考资料指南）

美国紫菀

小须芒草

斑茎泽兰

蝴蝶草

工作表

雨水花园设计注意事项

设计雨水花园时，可能需要考虑以下几点：

- 选择各种各样的植物时，包括灌木、花和草，要注意颜色、高度和质地上的多样性。
- 想象一下雨水花园全年的样子——丛生的草会在整个冬天保持它们的形状，许多类型的灌木在寒冷的月份会长出醒目的红色枝条。
- 考虑一下庭院的现有景观，以及周围的环境。
- 如果雨水花园在道路附近，考虑一下视线设计效果。

专业词汇

tolerate/ˈtɒləreɪt/*vt.* 忍受；默许；宽恕

alternate/ˈɔːltəneɪt/*adj.* 交替的；轮流的

fern/fɜːn/*n.* [植] 蕨；[植] 蕨类植物

criteria/kraɪˈtɪərɪə/*n.* 标准，条件

Designing Your Rain Garden

Take a minute and draw out a rough sketch of your ideal rain garden:

Building Your Rain Garden

Tools Required:

- Shovels
- Ground paint or string

Materials Required:

- Mulch *(refer to page 58 for quantity calculation)*
- Sand/compost mix *(refer to page 58 for quantity calculation)*
- A collection of small rocks and stones
- A variety of native plants *(refer to page 62 for quantity calculation)*

Steps:

1 Digging your rain garden

Before you even pick up a shovel, mark out your rain garden area to make sure you are happy with the size and how it will look. Ground paint, or string tied to wooden stakes are both easy ways to mark the area, and let you make changes if you would like. You should also mark out the area where water will get to your rain garden, and where the water overflow area will be (refer to step 3 for more information).

Once you are happy with the layout for your new garden, it's time to start digging. Rain gardens should be dug to a depth of approximately 85 centimetres (or deeper if you're building a small rain garden – refer to page *50*). The ground

设计雨水花园

花一分钟时间，画一幅你理想中的雨水花园草图：

建造雨水花园

所需的工具

- 铲子
- 地面油漆或细绳

所需的材料

- 护根覆盖物（请参阅第 59 页的数量计算）
- 沙/堆肥混合料（请参阅第 59 页的数量计算）
- 一堆小块岩石和石头
- 各种本土植物（请参阅第 63 页的数量计算）

步骤

1. 开挖

在拿起铲子之前，先将雨水花园的位置标记出来，确保对它的大小和设计效果满意。使用地面涂料，或者将细绳绑在木桩上，这都是标记雨水花园范围的简单方法，如果愿意也可以随时更改。还应该标记出雨水花园进水口的位置，以及雨水花园溢水口的位置（更多信息请参见步骤 3）

如果对新花园的布局感到满意，就可以开挖了。雨水花园应该挖到大约 85 厘米深（如果你正在建造一个小的雨水花园可以挖得更深，请参阅第 51 页）。雨水花园下面的岩土层不应压实，所以人应尽量站在花园区域之外，并避免在挖掘区域内放置任何重物或机械设备。

专业词汇

string/strɪŋ/*n.* 线，弦，细绳

wooden stake 木桩

layout/ˈleɪaʊt/*n.* 布局；设计；安排；陈列

under a rain garden should not be packed down, so try to stand outside of the garden area as much as you can, and avoid placing any heavy instruments or machinery inside the digging area.

2 Soil for your rain garden

Once you have finished digging your garden area, it's time to fill it back up with rain garden soil mix. This is one of the major differences between building a rain garden and building a regular garden; rain gardens require a special soil mix to work properly. This soil mix usually contains a combination of compost and sand and is designed to allow water to infiltrate easily and quickly.

Soil mix should be added to your excavated rain garden area to a depth of approximately 60 centimetres (this will leave approximately 25 centimetres for plants, mulch and water pooling - remember, rain gardens should be sunken into the ground to allow rain water to collect and filter into the soil). When filling with soil, lightly tamp down the area with your feet every 20 centimetres or so to help the soil settle.

3 Getting water to the rain garden and providing an area for overflow

You need to create the inlet (area for water to enter the rain garden) and the overflow (area for water to exit the rain garden in case of overfilling).

Creating an Inlet

Water can be carried to your rain garden in a variety of different ways. It can be carried through an extension on the end of your downspout, across a landscaped area, through an infiltration trench lined with plants or decorative rock (refer to the "Soakaways" chapter, page 64), or through the overflow pipe from your rain barrel (refer to the "Rain Barrds" chapter, page 30). A few small stones should be placed around the spot where water will enter the rain garden to slow the water and protect the soil from being washed away.

Creating an Overflow

During most of the year, all the water that flows into your rain garden will soak into the ground. Occasionally, when the ground is wet and a big storm occurs, the garden can fill up and overflow. A rain garden needs to have an overflow area to let excess water leave the rain garden. The overflow should be located on the downhill side of the garden, directing excess water towards a second garden or an open area of lawn. As you did with the inlet, you should also place a few small stones around

Anatomy of a Rain Garden

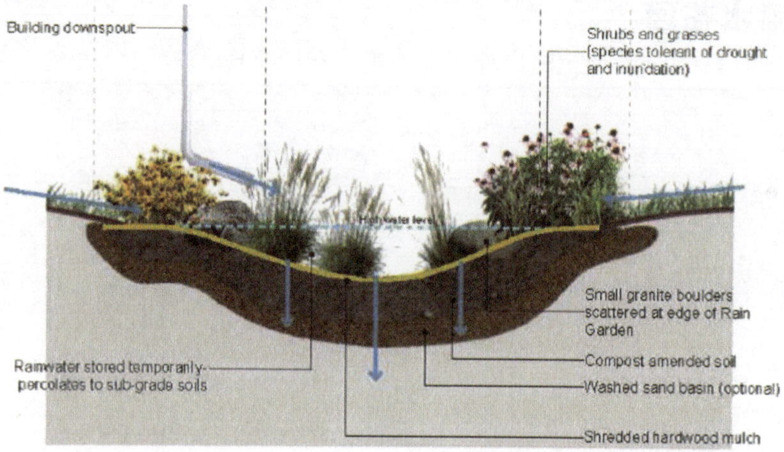

the overflow area to slow the water as it exits and protect the soil from being washed away.

4 Planting your rain garden

Planting your rain garden is the final step. Refer back to your original design plan on page 54 when planting. It's best to position all of your plants in the garden before removing them from their containers to make sure you like the layout. Now you can start planting.

Turn the plant's container upside down, give the bottom of the pot a sharp rap, and catch the root ball as the plant slides out from the container. Tease the roots out with your hands – don't worry if you break a few of the roots.

Make sure you do not plant the new plant too deep into the soil. The depth should be equal to the depth in the container. Also, make sure you consider the plant's mature size when spacing them – you don't want to overcrowd your garden. Press the soil firmly around its base and water well. Plants will need to be watered every few days until established (about 4 weeks). Once established, the plants will not need to be watered regularly, except for during times of drought.

Once all of your plants are in, cover your new garden with a 5-10 centimetres layer of mulch. Mulch helps to keep out weeds, resists drought and gives the garden a tidy appearance. Mulch is especially important for rain gardens as it also helps to absorb and filter the rain water. Your layer of mulch should not be more than 10 centimetres deep.

2. 雨水花园的土壤

开挖结束后，应用土壤混合物对雨水花园进行回填。这是建造雨水花园和普通花园的主要区别之一。雨水花园需要一种特殊的土壤混合物才能很好地工作。这种土壤混合物通常是堆肥和沙子的混合物，目的是让雨水能够快速、容易地入渗。

雨水花园土壤混合物的回填深度大约 60 厘米（留下大约 25 厘米深度用于植物生长、护根铺设和储水——记住，雨水花园应是下沉式的，以便雨水收集和过滤）。填土时，每隔 20 厘米左右用脚轻轻夯实，以使土壤稳定。

3. 为雨水花园设置引水区和溢流区

建造雨水花园需要设置引水区（水进入雨水花园的区域）和溢流区（水溢出雨水花园的区域，以防积水过多）

设置引水区

水可以以各种不同的方式进入雨水花园。可以通过落水管末端的延伸装置，穿过景观区，再穿过有植物或装饰性岩石的渗透沟渠进入雨水花园（参阅第 65 页“渗水坑”部分），或者通过雨水桶的溢流管（参见第 31 页“雨水桶”部分）。在雨水进入雨水花园的地方应放置一些小石头，以减缓水流速度，保护土壤不被冲走。

设置溢流区

在一年的大部分时间里，所有流入雨水花园的雨水都会渗入地下。有时，当雨水花园的土壤湿润而又下大雨时，雨水会灌满花园并溢出。雨水花园需要设置溢流区，让多余的水流出雨水花园。溢流口应位于花园的下坡一侧，将多余的水引向第二个花园或一片开阔的草坪。像建造引水区一样，也应在溢流口周围放置一些小石头，以减缓水流速度，保护土壤不被冲走。

4. 种植

种植是建造雨水花园的最后一步。种植时请参阅第 55 页的初步设计。最好把所有的植物都在雨水花园中放置好，然后再把它们从花盆中拿出来，以确保你喜欢这样的种植布局。现在，可以开始种植了。

把花盆倒过来，在花盆底部狠狠敲一下，当植物从花盆中滑出来时抓住根部。用手梳理植物的根——如果弄断了一些根，也不要担心。

确保不要种得太深，种植深度应该与在花盆中的种植深度相当。另外，确保在种植的时候考虑了植物长成后的尺寸——你肯定不想让花园太拥挤。把植物根部周围的土壤压实、浇透。植物需要每隔几天浇一次水，直到成长稳定（大约 4 周）。一旦稳定，这些植物将不再需要定期浇水，除非在干旱时期。

所有的植物都种好后，用 5～10 厘米厚的护根覆盖物覆盖新花园。护根覆盖物可以帮助除草、抗旱，使花园外观整洁。覆盖物对于雨水花园尤其重要，因为它也有助于吸收和过滤雨水。覆盖层的厚度不应超过 10 厘米。

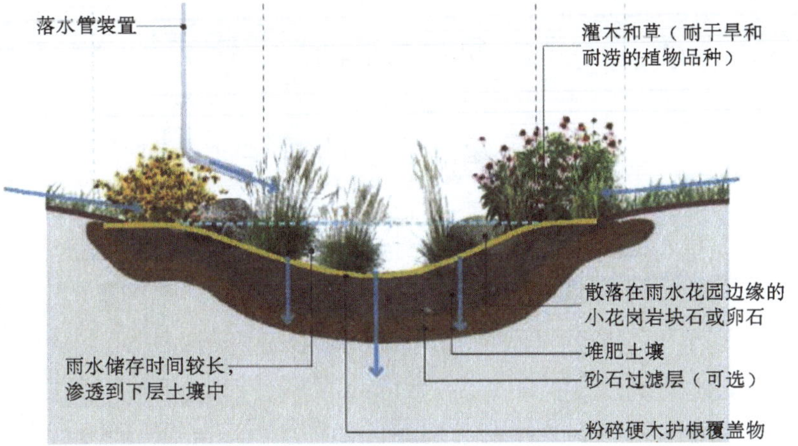

雨水花园剖面图

专业词汇

excavate/ˈekskəveɪt/*v.* 发掘；挖掘

occasionally/əˈkeɪʒnəli/*adv.* 偶尔；间或

mature/məˈtʃʊə(r)/*adj.* 成熟的

established/ɪˈstæblɪʃt/*adj.* 确定的；已制定的，已建立的

How Much Soil Should I Order?

What is your soil type?

Clay soil

Replace all the excavated soil with a rain garden soil mix of:

- About 60% sand and 40% compost

_______metres (garden length)
x _______ metres (garden width)
x 0.6 m (soil depth)
= _______ m^3 of soil mix

For volume of compost needed:
_______ m^3 of soil mix
x 0.4
= _______ m^3 of compost

For volume of sand needed:
_______ m^3 of soil mix
x 0.6
= _______ m^3 of sand

Sandy or Loam soil

Set aside the excavated soil and mix with compost for refilling the rain garden. The mixture should be:

- About 60% existing soil and 40% compost

_______metres (garden length)
x _______ metres (garden width)
x 0.6 m (soil depth)
= _______ m^3 of soil mix

For volume of compost needed:
_______ m^3 of soil mix
x 0.4
= _______ m^3 of compost

For volume of existing soil* needed:
_______ m^3 of soil mix
x 0.6
= _______ m^3 of existing soil*

**Remember, this is the soil you've saved from excavating the garden area. You don't need to purchase more.*

You should be able to find sand and compost at your local garden centre. Many centres will even create the custom mix required if you ask.

HOW MUCH MULCH SHOULD I ORDER?

The calculation to determine how much mulch you will need to fill your new rain garden is similar to the calculation you just did (to the left) to determine how much soil mix you will need.

___ metres *(garden length)* x ____ metres *(garden width)*
x ______ metres *(mulch depth – should be around 0.1m)*

= ______ m^3 of mulch

CONVERSION OF CUBIC METRES TO CUBIC YARDS

Many garden centres still price their bulk aggregate (soil, sand, compost, mulch) in cubic yards. It is easy to convert your measurements:

1 cubic metre = 1.3 cubic yards

So to covert any measurements in cubic metres to cubic yards, simply multiply by 1.3.

应该订购多少土壤?

你的土壤类型是什么?

黏质土壤

将所有挖掘出来的土壤替换为雨水花园专用土壤混合物：大约60%的沙子和40%的堆肥。

____米（花园长度）×____米（花园宽度）×0.6米（土壤厚度）=____m^3（土壤混合物）

所需堆肥量：____m^3（土壤混合物）×0.4=____m^3（堆肥量）

所需沙量：____m^3（土壤混合物）×0.6=____m^3（沙子量）

砂土或壤土

将挖出的泥土放在一边，与堆肥混合，用于回填雨水花园。混合物为：大约60%的现有土壤和40%的堆肥。

____米（花园长度）×____米（花园宽度）×0.6米（土壤厚度）=____m^3

所需堆肥量：____m^3（土壤混合物）×0.4=____m^3（堆肥量）

所需现有土壤体积：____m^3（土壤混合物）×0.6=____m^3（现有土壤）

*记住，这是你在花园中挖出来的土壤，你不需要再购买。

在当地的园艺中心应该能够找到所需的沙子和堆肥。如果需要，很多园艺中心甚至可以帮你混合好。

应该订购多少护根覆盖物?

需要多少护根覆盖物来满足雨水花园需要，计算方法与刚刚做的（左边）土壤混合物的计算方法相似。

________m（花园长度）×________m（花园宽度）×________m（护根覆盖物厚度约0.1m）=________m^3（护根覆盖物量）。

将立方米换算为立方码

许多园艺中心仍然以立方码为单位价格出售散装材料（土壤、沙子、堆肥、护根物）。测量结果很容易转换：

1立方米=1.3立方码

因此，要将任何以立方米为单位的测量数据转换成立方码，只需乘以1.3。

专业词汇

purchase /pɜːtʃəs/ *vt.* 购买

Maintaining Your Rain Garden

A well-designed rain garden needs minimum care. However, just like any garden, it requires some maintenance to look good and function properly. There are five things to remember when maintaining your rain garden over the years.

Mulching

Every year check the mulch layer and, if needed, apply more to maintain a layer that is about 10 centimetres thick. Mulch can be applied any time of the year, but it's usually easiest to remember in spring at the start of the growing season.

Erosion and Debris

Water flowing into the rain garden can carry with it various types of debris that can clog the soil mix and slow drainage. Fast flowing water can also slowly eat away at your soil layer, washing it away and damaging your garden.

To guard against soil erosion and debris build-up:

- Check the rain garden for areas of exposed soil and re-mulch as needed. Mulch helps to absorb water and protect the soil underneath.
- Maintain a healthy cover of plants, and replace any dead plants to fill in holes.
- Maintain a cover of decorative rock around the inlet and overflow area to protect the soil.
- At least twice a year, check around the inlet and overflow areas for debris build-up such as leaves, sticks, and other items. Keep these areas clear to help your garden absorb rain water.

Fertilizing

Do not apply fertilizers to your rain garden. The rain garden soil mix provides plenty of nutrients and the native plants in your garden are well suited to local growing conditions, so extra fertilizing is not needed.

Weeding

Rain gardens will still soak up and filter rain water even if they are full of weeds. However, the plants in the rain garden will likely not grow as well with all the competition and your rain garden may not be as attractive. Soils in rain gardens have good structure, so weeds should be easy to pull by hand, especially in the spring when the soil is moist and the weeds are small.

Watering

For the first year most plants need deep watering during the dry summer season to establish healthy root systems. After two or three years the native plants in your rain garden will need little or no watering. However, watering may be necessary during prolonged dry periods even if plants are established.

WATERING TIPS

- Water deeply, but infrequently, so that the top 15 to 30 centimetres of the root zone is moist. To know if you're applying enough water, dig down 20 to 40 centimetres off to the side of the plant a few hours after watering—don't disturb the roots. If the ground it wet here, you know you've applied enough water.
- Use soaker hoses to ensure water soaks directly into the soil and isn't lost to evaporation.
- Don't over water. Over watering can rot plant roots and soggy soil reduces oxygen needed for healthy soil and plants.
- Water in the morning or evening to reduce evaporation during warm mid-day temperatures.

WHEN TO SEEK PROFESSIONAL ADVICE

If your soil is high in clay *(refer to page 14 for your soil analysis results)*, it can be challenging to create a highly functioning rain garden. You may want to consult with a landscape professional before proceeding for advice on dealing with your specific soil conditions.

Refer to page 106 and look under "Rain Gardens" for more information.

雨水花园维护

精心设计的雨水花园需要精心的照料。就像任何花园一样，雨水花园需要一些维护来保持它的外观和良好运行。经过多年实践，维护雨水花园要记住五件事。

覆盖

每年都要检查护根覆盖层，如果需要的话，增加覆盖物，使覆盖层保持在大概 10 厘米的厚度。覆盖物可以在一年中的任何时候添加，但是通常在春天这个万物开始生长的季节最容易被记住。

侵蚀和杂物堆积

流入雨水花园的水会携带各种各样的杂物，这些杂物会堵塞土壤混合物的孔隙并导致渗水缓慢。快速流动的水也会慢慢侵蚀土壤，引起水土流失，破坏雨水花园。

防止土壤侵蚀和杂物堆积：

- 检查雨水花园是否有暴露的土壤，必要时重新覆盖。护根覆盖物有助于吸收水分，保护土壤。
- 保持植物的健康生长，并替换枯死的植物。
- 在入水口和溢流口周围设装饰性岩石覆盖，以保护土壤。
- 每年至少两次，检查入水口和溢流口是否有树叶、树枝等杂物堆积。保持这些区域的清洁，帮助花园消纳雨水。

施肥

不要在雨水花园里施肥。雨水花园的土壤混合物提供了大量的营养，花园中的本土植物非常适合当地的生长条件，所以不需要额外施肥。

除草

雨水花园即使长满了杂草也仍然会吸收和过滤雨水。然而，杂草不仅会影响雨水花园中的植物生长，也会影响整体美观。雨水花园的土壤具有良好的结构，所以杂草应该很容易用手拔除，特别是在春天土壤潮湿、杂草比较小的时候。

浇水

第一年，大多数植物在干燥的夏季需要深度浇水以长成健康的根系。两三年后，雨水花园中的本土植物就不需要浇水了。然而，即使植物已经成长稳定，长时间的干旱期也是有必要浇水的。

浇水技巧

- 深度浇水，但并不频繁，这样根系生长层的上部 15～30 厘米会保持湿润。要想知道是否浇了足够的水，可以在浇水几小时后，在植物的旁边挖 20～40 厘米——注意不要破坏植物的根部。如果挖出的土壤是湿的，那你浇的水量就够了。
- 使用渗水软管，确保水直接渗入土壤，不会蒸发。
- 不要过度浇水。过度浇水会使植物根部腐烂，浸水的土壤会减少健康土壤和植物所需的氧气。
- 早上或晚上浇水，以减少正午高温下的蒸发。

何时寻求专业意见？

如果你的土壤是高黏性土（参见第 15 页的土壤分析结果），建造一个功能强大的雨水花园是很有挑战性的。你可以先咨询一下专业的景观设计师，然后再去咨询如何改良土壤。

请参阅第 107 页，在“雨水花园”下寻找更多信息。

专业词汇

soil erosion　水土流失

nutrient/ˈnjuːtriənt/*n*. 营养物；滋养物

infrequent/ɪnˈfriːkwənt/*adj*. 罕见的；稀少的；珍贵的；不频发的

disturb/dɪˈstɜːb/*vi*. 打扰；妨碍

evaporation/ɪˌvæpəˈreɪʃn/*n*. 蒸发；消失

soggy/ˈsɒgi/*adj*. 浸水的；透湿的

Rain Garden Costs

Use the cost sheet below to estimate the costs for installing your rain garden. Pricing provided is an estimate only – it varies depending on area, time of year, and sources. However, on average, rain gardens typically cost approximately $15 per square metre of drainage area, or $90 per square metre of garden area.

Garden Area:	______ m^2	
Total Depth:	______ m	
Compost:	______ m^3 x $65	$______ (refer to page 58 for quantity)
Sand:	______ m^3 x $65	$______ (refer to page 58 for quantity)
Mulch:	______ m^3 x $80	$______ (refer to page 58 for quantity)
Delivery:		$______ ($75.00 is a good estimate for delivery costs)
Plants*:	10 x ______ m^2 x $5	$______
River rocks**:	______ bags x $8	$______
Help of friends and family:		priceless
Total:		**$______**

*As a general rule, you will need approximately 10 plants per square metre of garden space. This will vary depending on species used, so use your judgement when purchasing. (Pricing is for a 10 centimetre pot)

**The quantity of river rocks you will require depends on where you will be using them. Use your best estimate here. (Pricing is for an 18 kilogram bag)

雨水花园的成本

使用下面的成本计算表来估算建造雨水花园的成本。所提供的价格只是估算结果——它取决于所在地区、时间和材料来源。然而，平均而言，典型的雨水花园约合每平方米汇水面积 15 加元，或每平方米花园面积 90 加元（按 2019 年 1 月 31 日汇率，约为 459 元人民币）。

花园面积：__________ m^2	
总深度：__________ m	
堆肥：__________ m^3 ×$ 65	$______（数量参见第 59 页）
沙：__________ m^3 ×$ 65	$______（数量参见第 59 页）
护根覆盖物：______ m^3 ×$ 80	$______（数量参见第 59 页）
送货费：__________	$______（75.00 加元是很合适的送货成本）
植物*：10×______ m^2 ×$ 5	$______
河流岩石**：______ bags×$ 8	$______
朋友和家人的帮助：	无价
合计：	$______

* 一般来说，每平方米的花园空间大约需要 10 棵植物。这将根据使用的植物种类有所不同，所以在购买时请根据需要来判断。（售价为 10 厘米花盆所种植物的价格）

** 需要的河流岩石数量取决于在哪里使用它们，这里用估计量即可。（售价为 18 公斤袋装的价格）

Soakaways

A soakaway is a small underground reservoir that collects rainwater runoff from your downspout and allows it to slowly soak into the surrounding soil. Soakaways can be rectangular or circular. They are usually lined with landscape fabric, filled with clean gravel, and camouflaged with plants or decorative rocks. Often, soakaways are designed to be installed along with a rain barrel or rain garden. (See "Connecting the Drops" for examples on how this is done).

Advantages and Disadvantages of Soakaways

Advantages:

- Good alternative solution to a rain garden, when limited by space on your property.
- Used for carrying water from one location to another and allowing water to soak back into the ground.
- Require very little maintenance over the years.
- Inexpensive.

Disadvantages:

- Will not function properly if clogged by leaf litter and soil.
- Require a fair amount of digging to reach appropriate depth.

RAINGARDENS AND SOAKAWAYS.......WHAT'S THE DIFFERENCE?

Rain gardens and soakaways are actually quite similar. They both accept water from your downspout and allow it to soak into the ground. The main difference is that rain gardens are filled with a sand/compost mix and soakaways are filled with stone. Like a rain garden, soakaways have the option of a top layer of soil for plants, or you can turn them into a rock garden feature. Soakaways can fit in small and narrow spaces, where as rain gardens usually need larger areas.

ARE SOAKAWAYS RIGHT FOR ME?

	YES	NO
Do you have a water source to feed your soakaway, such as a disconnected downspout, or rain barrel?	☐	☐
Do you have limited space for a rain garden?	☐	☐
Do you like low maintenance landscaping?	☐	☐
Are you willing to dig down into your property?	☐	☐

If you answered 'YES' to all of the above, you are ready to install your soakaway! Read on to learn how.

渗水坑

渗水坑是一个小型的地下水库，收集从落水管流出的雨水，并使雨水慢慢渗入周围的土壤。渗水坑可以是方形的，也可以是圆形的。它们通常内衬土工织物，用干净的砂或砾石填充，用植物或装饰性的岩石覆盖。通常可以把渗水坑与雨水桶或雨水花园设计为一个整体。（请参阅“雨水的连通路径”部分的内容，以了解如何做到这一点。）

渗水坑的优点和缺点

优点：

- 当庭院空间有限时，渗水坑是替代雨水花园的好方案。
- 可以将雨水从一个地方输送到另一个地方，输送过程中雨水可以渗入地下。
- 长期的维护成本很低。
- 造价低。

缺点：

- 如果被落叶和土壤颗粒堵塞，渗水坑将不能正常工作。
- 需要大量的土方开挖才能达到合适的深度。

渗水坑和雨水花园有什么区别？

实际上，雨水花园和渗水坑非常相似，它们都收集房屋落水管里的雨水，并使雨水渗入地下。最主要的不同之处在于，雨水花园中填充的是沙子或堆肥混合物，而渗水坑中填充的是石头。渗水坑也可以像雨水花园一样，将上层设置为土壤以方便种植植物，或者全部由岩石填充，成为石景花园。渗水坑能适应小而狭窄的空间，不像雨水花园那样需要较大的区域空间。

渗水坑适合我家的庭院吗？

	是	否
• 是否有充足的水源供给渗水坑，比如断开的落水管或者雨水桶？	□	□
• 是否有充足的空间修建雨水花园？	□	□
• 是否喜欢维护费用低的雨水景观类型？	□	□
• 是否愿意在庭院中开挖土方？	□	□

对于上述所有问题，如果都回答“是”，那么你已经准备好建造渗水坑了！

继续往下阅读，学习具体做法。

专业词汇

reservoir/ˈrezəvwɑː(r)/*n.* 水库，蓄水池

soak/səʊk/*vt.* 吸收，吸入；沉浸在（工作或学习中）；使…上下湿透；*vi.* 浸泡；渗透

surrounding/səˈraʊndɪŋ/*adj.* 周围的，附近的

rectangular/rekˈtæŋgjələ(r)/*adj.* 长方形的，矩形的，成直角的

circular/ˈsɜːkjələ(r)/*adj.* 循环的，圆形的

fabric/ˈfæbrɪk/*n.* 织物，布，组织，构造

gravel/ˈgrævl/碎石，砾砂

camouflage/ˈkæməflɑːʒ/*n.* 伪装，掩饰；*vt.* 伪装，掩饰；*vi.* 伪装起来

decorative/ˈdekərətɪv/*adj.* 装饰性的；装潢用的

alternative/ɔːlˈtɜːnətɪv/*adj.* 供选择的；选择性的；交替的；*n.* 二者择一；供替代的选择

function/ˈfʌŋkʃn/*vi.* 运行；活动；行使职责；*n.* 功能；[数] 函数；职责；盛大的集会

clog/klɒg/*v.* 阻塞；障碍；*n.* 障碍；木底鞋

leaf/liːf/*n.* 叶子；（书籍等的）一叶；一张；扇叶

leaf litter　落叶层

dig/dɪg/*vt.* 挖，掘；探究；*vi.* 挖掘；*n.* 戳，刺；挖苦

reach/riːtʃ/*vi.* 达到；延伸；伸出手；传开；*vt.* 达到；影响；抵达；伸出

appropriate/əˈprəʊprɪət/*adj.* 适当的；恰当的；合适的

quite/kwaɪt/*adv.* 很；相当；完全

feature/ˈfiːtʃə(r)/*n.* 特色，特征；容貌；特写或专题节目；*vi.* 起重要作用；*vt.* 特写，以…为特色；由…主演

fit in　适应，适合；装配好；找时间做

source/sɔːs/*n.* 来源；水源；原始资料

feed/fiːd/*vt.* 喂养；供给；放牧；抚养（家庭等）；*vi.* 吃东西，流入

Which Soakaway is Best Suited for Your Property?

Infiltration Trench

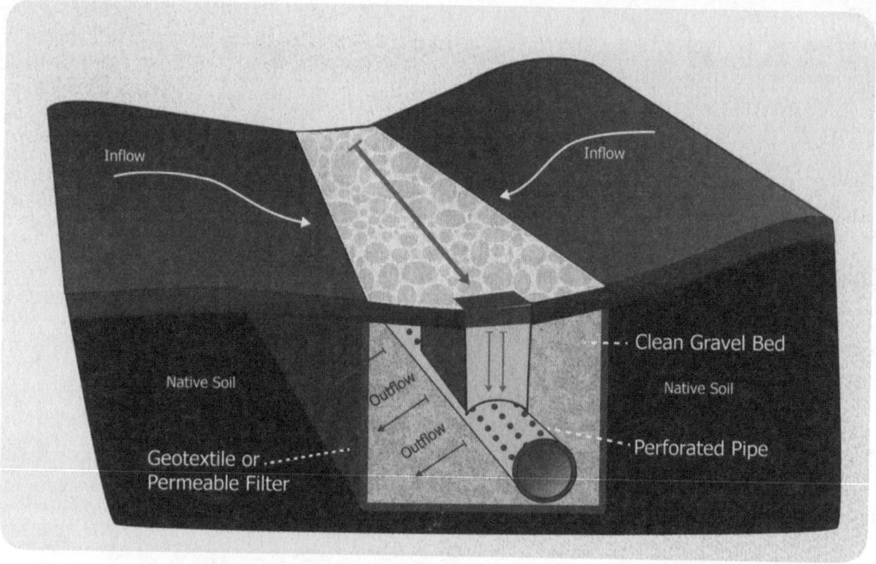

There are a variety of soakaway options to consider. In general they can be divided into two categories.

1. **Infiltration Trenches**
   An infiltration trench is a linear soakaway – basically a long narrow underground reservoir filled with stones. It is designed to carry water from one area to another while allowing it to soak into the ground. It is perfect for narrow strips of land between suburban driveways.

   Infiltration trenches can range in size from 30 centimetres to 2 metres deep, up to 1 metre wide, and can be any length depending on the amount of area you have available. They must be slightly sloped in order to allow water to flow from one end to the other. They are used to direct water towards permeable surfaces or other collection systems such as soakaway pits or rain gardens (page 46).

Soakaway Pit

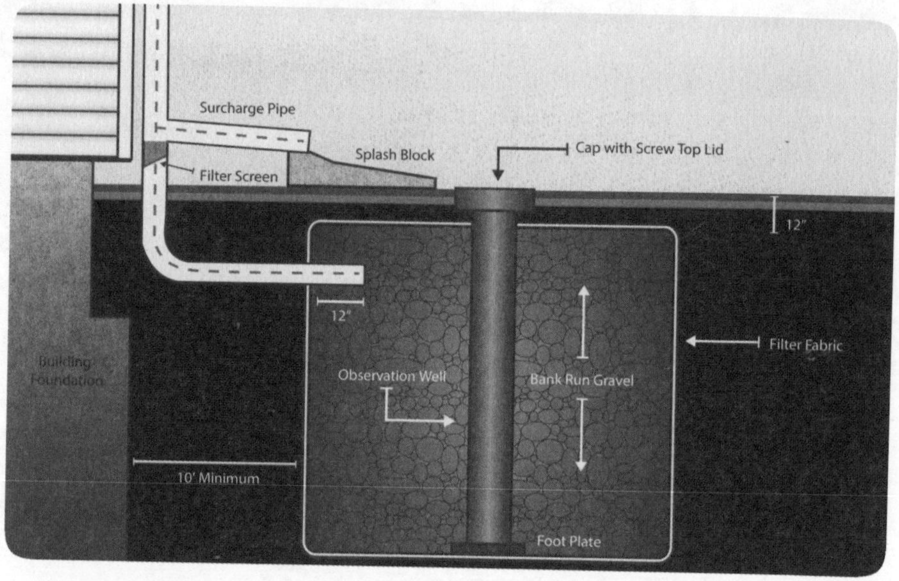

2. **Soakaway Pits**
   Soakaway pits are underground pits. They are filled with stones and collect water, allowing it to slowly soak into the surrounding soil. A disconnected downspout or infiltration trench can feed into the pit. You can cover your soakaway pit with grass to make it appear "invisible", or create a landscaped rock garden.

哪种类型的渗水坑最适合你的庭院？

有各种类型的渗水坑可供选择。一般来说，可分为两类。

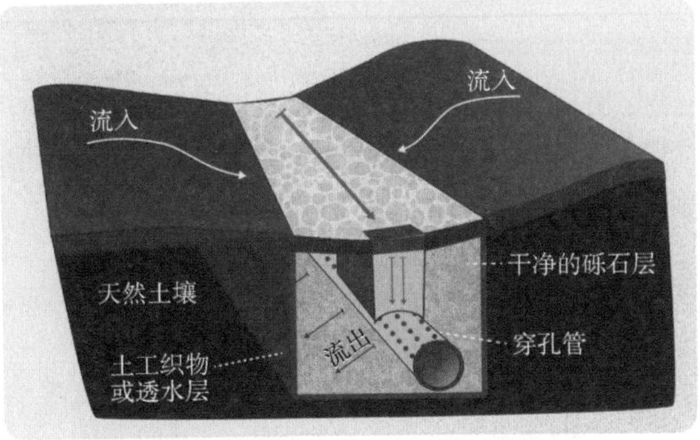

图 1　渗透沟

1. 渗透沟

渗透沟是一个线状的渗水坑——基本由石头填充，像一个狭长的地下水库。这样设计的目的是将雨水从一个区域输送到另一个区域，在输送的同时允许雨水渗入地下。对于庭院中行车道边狭长的地带而言，这是一种很好的选择。

渗透沟的深度从 30 厘米到 2 米不等，宽度可达 1 米，长度视庭院中的可用面积而定。为了让雨水从一端流向另一端，渗透沟在长度方向上必须稍微倾斜，这样才可以将水引向可渗透的地表或其他雨水收集系统，如渗透井或雨水花园（第 47 页）。

2. 渗水坑

在庭院中，渗水坑是位于地下的深坑。渗水坑中填满了石头和收集来的雨水，雨水可以慢慢地渗入周围的土壤中。一个断开的落水管或渗透沟都可以延伸至渗水坑。可以用植被覆盖渗水坑，让它在庭院中是“看不见的”，或者建造一个石景花园。

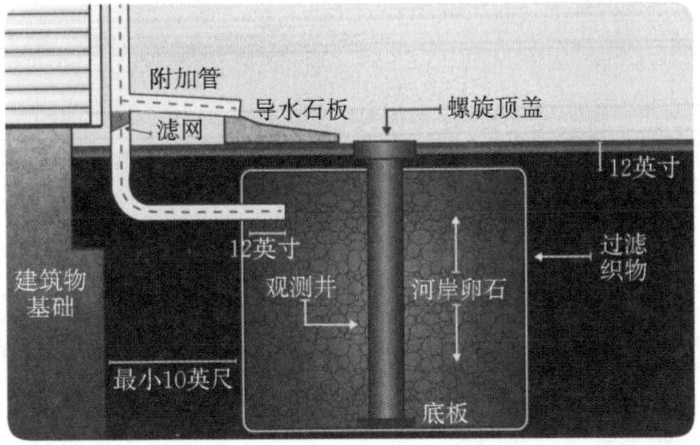

图 2　渗水坑

专业词汇

suited/ˈsjuːtɪd/*adj*. 合适的；*v*. 适合（suit 的过去分词）

category/ˈkætəgəri/*n*. 种类；分类

infiltration/ˌɪnfɪlˈtreɪʃn/*n*. 渗透；渗透物

trench/trentʃ/*n*. 沟，沟渠；战壕；堑壕；*vt*. 掘沟；*vi*. 挖战壕；侵害

linear/ˈlɪnɪə(r)/*adj*. 线的，线型的；直线的，线状的；长度的

strip/strɪp/*n*. 带；条状；脱衣舞；*vt*. 剥夺；剥去；脱去衣服；*vi*. 脱去衣服

suburban/səˈbɜːbən/*adj*. 郊区的，城郊的；土气的；见闻不广的；*n*. 郊区居民

driveway/ˈdraɪvweɪ/*n*. 车道

range/reɪndʒ/*n*. 范围；幅度；排；山脉；*vi*. 平行，列为一行；延伸；漫游；射程达到

range from　从…到…变动；从…到…范围；延伸；绵亘

centimetre/ˈsentɪmiːtə(r)/*n*. 厘米；公分

available/əˈveɪləbl/*adj*. 有效的，可得到的；可利用的；空闲的

slightly/ˈslaɪtli/*adv*. 些微地，轻微地；纤细地

direct/dɪˈrekt/*vt*. 管理；指挥；导演；指向；*vi*. 指…；*adj*. 直接的；直系的；亲身的；恰好的

permeable/ˈpɜːmɪəb(ə)l/*adj*. 能透过的；有渗透性的

inflow/ˈɪnfləʊ/*n*. 流入；流入物；*vi*. 流入；货币回笼

outflow/ˈaʊtfləʊ/*n*. 流出；流出量；流出物；*vi*. 流出

native/ˈnetɪv/*adj*. 本国的；土著的；天然的；与生俱来的；*n*. 本地人；土产；当地居民

gravel bed 砾石层

geotextile/ˌdʒiəʊˈtekstail/*n*. 土工布，隔泥纺织物料；土工织物

perforate/ˈpɜːfəreɪt/*vt*. 穿孔于，打孔穿透；在…上打齿孔；*vi*. 穿过，贯穿；穿孔

perforated pipe　多孔管，穿孔管

pit/pɪt/*n*. 矿井；深坑；陷阱；（物体或人体表面上的）凹陷；*vt*. 使竞争；窖藏；使凹下去

feed into 注入，流入；提供原料

grass/grɑːs/*n*. 草；草地，草坪；*vt*. 放牧；使…长满草；使…吃草；*vi*. 长草

invisible/ɪnˈvɪzəbl/*adj*. 无形的，看不见的；不显眼的，暗藏的

building foundation　建筑物基础

surcharge/sɜːˈtʃɑːdʒ/*n*. 超载；额外费；*vt*. 追加罚款；使…装载过多；使…负担过重

filter/ˈfɪltə(r)/*vi*. 过滤；渗入；慢慢传开；*n*. 滤波器；[化工] 过滤器；筛选；滤光器

filter screen 滤网

splash/splæʃ/*vt*. 溅，泼；用…使液体飞溅；*vi*. 溅湿；溅开；*n*. 飞溅的水；污点；卖弄

block/blɒk/*n*. 块；街区；大厦；障碍物；*vt*. 阻止；阻塞；限制；*adj*. 成批的，大块的；交…

cap/kæp/*n*. 盖；帽子；*vi*. 脱帽致意；*vt*. 覆盖；胜过；给…戴帽；加盖于

screw/skruː/*vt*. 旋，拧；压榨；强迫；*vi*. 转动，拧；*n*. 螺旋；螺丝钉；吝啬鬼

lid/lɪd/*n*. 帽子；眼睑；限制；*vt*. 给…盖盖子

filter fabric　过滤织物

plate/pleɪt/*n*. 碟；金属板；金属牌；感光底片；*vt*. 电镀；给…装甲

Getting Started

A soakaway must be carefully designed and maintained to work properly. Poorly installed or improperly located soakaways can fail, and will not achieve the desired results. Before starting any soakaway project, it is important to perform the following steps as part of your site assessment.

Soils

Porous soil with lots of sand is best suited for soakaways. It allows the water to soak into the ground quite quickly. Clay soil allows water to soak into the ground at a much slower rate. If you are dealing with heavy clay soils you may need to dig further down and replace the soil with a sand-compost mixture. Refer to page 14 for instructions on determining your soil type, and calculating your infiltration rate.

Location

Finding the correct location for your soakaway is extremely important. You can be restricted by a number of factors such as soil type, slope of land, distance to the foundation of your home and obstructions like large tree roots and underground utilities. Before digging, we advise you obtain the proper dig permits.

Soakaway pits and infiltration trenches work best as collection areas for stormwater that runs out of your roof's downspouts. When you are considering where a soakaway could be installed on your property, keep in mind the following:

Space requirements

Infiltration trenches are usually between 30 centimetres and 1 metre deep, can be as wide as 1 metre and as long as your property allows. Soakaway pits can be designed to fit in small spaces, approximately 1 metre squared by 1 metre deep.

Proximity to Other Buildings

To prevent water from coming into contact with your foundation and possibly causing basement flooding, soakaways should be at least 3 metres away from yours or any neighbouring foundations.

Slope of Your Property

Properties with more than a 5% grade do not usually allow water to infiltrate into the ground and therefore are not recommended for soakaway projects. Refer to page 12 to calculate the slope of your property.

Drainage Problems

If you know of areas on your property that are prone to water pooling, a soakaway is not a good option. Water may be pooling because of a high water table, or because of compacted or heavy clay soil. However, if it is due to compacted or clay soil, you can potentially amend the soil to make it suitable for a soakaway. Refer to page 14 for calculating your soil infiltration rate.

Overflow

Like any structure accepting water, you should always install a drain or overflow in case of very large storm events. If the site permits, direct the overflow towards another stormwater management project such as a second soakaway or a rain garden. If this is not possible, have the overflow directed to a landscaped or grassy area downhill from the foundation of the home.

准备开始

渗水坑必须经过精心设计和维护才能正常工作。施工不当或选址不当都会导致渗水坑不能正常工作，而且达不到预期的效果。在渗水坑施工之前，作为现场评估的一部分，执行以下步骤是很重要的。

土壤

含有大量沙子的多孔隙土壤最适合于建设渗水坑，它能让水很快地渗入地下。黏土的渗透速度较慢，如果遇到重黏度的土壤，需要全部挖除，用沙土混合物置换。确定土壤类型和计算渗透率的有关说明，请参阅第 15 页。

位置

找到合适的位置是非常重要的，这可能会受到很多因素的限制，比如土壤类型、地面坡度、距离房屋地基的距离，以及大树根和地下设施等障碍物。在开挖之前，建议先获得开挖许可。

当雨水从屋顶的落水管流出时，渗水坑和渗透沟是收集雨水的好地方。当考虑在庭院的某个地方建设渗水坑时，请注意以下几点：

空间需求

渗透沟通常深 30 厘米到 1 米，宽度可达 1 米，长度视庭院条件确定。渗水坑也可以设计成 1 米见方的小空间。

临近其他建筑物

为了防止雨水浸泡地基，甚至淹没地下室，渗水坑距离房屋或任何邻近建筑物的地基至少 3 米。

庭院地面坡度

地面坡度超过 5%，通常不利于雨水下渗，因此不推荐修建渗水坑。请参阅第 13 页，计算庭院地面坡度。

排水问题

如果发现庭院中有容易积水的地方，那么也不适合修建渗水坑。积水的原因可能是由于地下水位高，或是土壤被压实，或是存在重黏土等。如果是由于土壤被压实或存在黏性土壤，可以对土壤进行改良，使其适合于修建渗水坑。请参阅第 15 页，计算土壤的渗透率。

溢流

像任何蓄水设施的结构一样，为应对强降雨事件，渗水坑应该设置排泄或溢流设施。如果场地允许，可将溢流出来的水导流到另一个雨水管理项目，如第二个渗水坑或雨水花园。如果没有其他雨水管理项目，可将溢流出来的水直接沿房屋基础下坡方向，引向雨水景观或草坪区。

专业词汇

getting started 入门指南，准备开始

properly/ˈprɒpəli/*adv.* 适当地；正确地；恰当地

project/prɒdʒekt/*n.* 工程；计划；*vi.* 设计；计划；表达；投射；*vt.* 设计；计划；发射；放映

perform/pəˈfɔːm/*vi.* 执行，机器运转；表演；*vt.* 执行；完成；演奏

site/saɪt/*n.* 地点；位置；场所；*vt.* 设置；为…选址

assessment/əˈsesmənt/*n.* 评定；估价

porous/ˈpɔːrəs/*adj.* 多孔渗水的；能渗透的；有气孔的

clay/kleɪ/*n.* ［土壤］黏土；泥土；肉体；似黏土的东西；*vt.* 用黏土处理

mixture/ˈmɪkstʃə(r)/*n.* 混合；混合物；混合剂

refer to 参考；涉及；指的是；适用于

restrict/rɪˈstrɪkt/*vt.* 限制；约束；限定

obstruction/əbˈstrʌkʃən/*n.* 障碍；阻碍；妨碍

utility/juːˈtɪləti/*n.* 实用；效用；公共设施；功用；*adj.* 实用的；通用的；有多种用途的

permit/pəˈmɪt/*n.* 许可证，执照；*vi.* 许可；允许；*vt.* 许可

permits 许可，许可证

stormwater/stɔːmˈwɔːtə/*n.* 雨水；暴雨水

approximately/əˈprɒksɪmətli/*adv.* 大约，近似地；近于

squared/skweəd/*adj.* 方格的；成正方形的；*v.* 使成正方形；求平方值（square 的 ing 形式）

proximity/prɒkˈsɪməti/*n.* 亲近；接近；［数］邻近

basement/ˈbeɪsmənt/*n.* 地下室；地窖

flooding/ˈflʌdɪŋ/*n.* 泛滥；产后出血

neighbouring/ˈneɪbərɪŋ/*adj.* 邻近的；附近的；接壤的

prone/prəʊn/*adj.* 俯卧的；有…倾向的；易于…的

prone to 有…倾向的

pool/puːl/*n.* 联营；撞球；水塘；共同资金；*vi.* 联营，合伙经营；*vt.* 合伙经营

pooling *n.* ［贸易］联营，合并；池化；*v.* 积水成池，淤积

amend/əˈmend/*vt.* 修改；改善；改进；*vi.* 改正，改善；改过自新

suitable/ˈsjuːtəbl/*adj.* 适当的；相配的 suitable for 适合…的，适合于…

overflow/ˌəʊvəˈfləʊ/*n.* 充满，洋溢；泛滥；超值；溢值；*vt.* 使溢出；使泛滥；*vi.* 溢出；泛滥；充溢

drain/dreɪn/*n.* 排水；下水道；排水管；*vt.* 喝光，耗尽；使流出；排掉水；*vi.* 排水；流干

event/ɪˈvent/*n.* 事件，大事；项目；结果

grassy/ˈgrɑːsi/*adj.* 长满草的；草绿色的

downhill/ˌdaʊnˈhɪl/*adv.* 下坡；向下；每况愈下；*adj.* 下坡的；容易的；*n.* 下坡；滑降

WORKSHEET - Sizing Your Soakaway

As with all of the projects presented in this guide, a small soakaway is better than nothing at all. But to get the most out of your soakaway, try to size it according to what your property can accommodate.

In order to determine the best size for your soakaway, you need to know how much water will be draining to it. The first step is to calculate the total drainage area (your roof area). To do this, you will have to measure the total width and length of your house, and then do the same for the garage (if applicable).

Example:

House: 8 metres x 17 metres = 136 m^2
+
Garage: 4 metres x 6 metres = 24m^2

Total roof area of house and garage:
= 160 m^2

Length of House: (A)____________ m

Width of House: (B)____________ m

Total Area of House (multiple A x B): (C) ____________ m^2

Length of Garage: (D)____________ m

Width of Garage: (E)____________ m

Total Area of Garage (multiply D x E): (F) ____________ m^2

Total Roof Area (add together C and F): (G) ____________ m^2

Next, to determine how much water, on average, drains to each downspout, divide the total roof area (G) by the number of downspouts on your house and garage. (G ÷ H = J)

Total number of downspouts: (H) ____________

(G ÷ H = J)

Total area draining to each downspout: (J) ____________ m^2

Finally, to determine the best size for your soakaway, divide this number by 10. (J ÷ 10 = K)

Your ideal soakaway size: (K) ____________ m^2

This is the recommended size for your soakaway. If you plan to make a larger soakaway, that's fine, but if you plan to make a smaller soakaway, you should dig down 10 – 20 centimetres deeper.

How Deep Do I Need to Go?

Referring back to page 14, you should already know what type of soil you have on your property.

Below we have provided dig depths for each soil type.

Soil Type	Recommended Depth of Soakaway (metres)
Sand	0.3
Sandy loam	0.5
Loam	0.75
Clay loam	1.0
Clay	1.25

IMPORTANT

Remember, all soakaways must be located at least 3 metres away from any building foundation.

工作表——确定渗水坑尺寸

与本手册中介绍的所有其他项目一样，有一个小型渗水坑总比什么都没有的好。为了最大限度地利用渗水坑，可以试着根据庭院的容纳能力来确定渗水坑的大小。

为了确定渗水坑最合适的尺寸，需要知道有多少水会被排入其中。第一步是计算总汇水面积（屋顶面积）。要做到这一点，必须测量房屋的总宽度和长度，然后测量车库的总宽度和长度（如果适用的话）。

例如：

房屋：8m×17m=136m²

＋

车库：4m×6m=24m²

房屋和车库的屋顶总面积：=160m²

房屋的长度：(A)__________ m

房屋的宽度：(B)__________ m

房屋的总面积（A×B）：(C)__________ m²

车库的长度：(D)__________ m

车库的宽度：(E)__________ m

车库的总面积（D×E）：(F)__________ m²

屋顶总面积（C 和 F 相加）：(G)__________ m²

接下来，确定每个落水管的平均汇水面积。用屋顶总面积（G）除以房屋和车库的落水管的数量。(G÷H=J)

落水管的总数：(H)__________

(G÷H=J)

每个落水管的汇水面积：(J)__________ m²

最后，把这个数字除以 10，得到渗水坑的最佳尺寸。

(J÷10=K)

理想的渗水坑面积：(K)__________ m²

这是渗水坑的推荐尺寸。如果打算做一个更大的渗水坑也可以，但是如果打算做一个更小的渗水坑，你应该往下深挖 10～20 厘米。

需要挖多深呢？

回到第 15 页，你应该已经知道庭院内是什么类型的土壤。

下表提供了每一种土壤类型的开挖深度推荐值。

土壤类型	渗水坑推荐深度/m
砂土	0.3
砂壤土	0.5
壤土	0.75
黏壤土	1.0
黏土	1.25

重要提示

记住，所有的渗水坑必须距离任何建筑物基础至少 3 米。

专业词汇

present/ˈprezənt/*vt.* 提出；介绍；呈现；赠送；*vi.* 举枪瞄准；*adj.* 现在的，出席的

presented *v.* 提出（present 的过去分词），呈递，提供

guide/gaɪd/*n.* 指南；向导；入门书；*vt.* 引导；带领；操纵；*vi.* 担任向导

better than nothing 聊胜于无，总比没有的好

nothing at all 一无所有，什么都不是

accommodate/əˈkɒmədeɪt/*vt.* 容纳；使适应；供应；调解；*vi.* 适应；调解

garage/ˈgærɑːdʒ/*n.* 车库；汽车修理厂；飞机库；*vt.* 把…送入车库

applicable/ˈæplɪkəbl/*adj.* 可适用的；可应用的；合适的

multiple/ˈmʌltɪpl/*adj.* 多重的；多样的；许多的；*n.* 倍数；［电］并联

sandy/ˈsændi/*adj.* 沙地的；多沙的；含沙的

located/ləʊˈketɪd/*adj.* 处于，位于；坐落的

WORKSHEET - Designing Your Soakaway

There are a couple things to consider when designing the overall look of your soakaway. If you wanted a rain garden but did not have the space, here is your chance to make your soakaway look like one.

- *If you decide to plant your soakaway, choose a variety of water loving flowers and grasses that vary in height, colour and texture (refer to page 52 for a list of appropriate plants).*
- *Like a rain garden, we ask you to consider the year round look of your soakaway. Grasses will hold their shape throughout the winter, but flowers will disappear causing bare spots in the colder months.*
- *Consider your home's existing landscape and how it will look against your current gardens, driveway, walkway and house.*
- *If the soakaway ends near the road, consider site lines and setbacks from the edge of your property.*

Take a minute and draw a rough sketch of what your ideal soakaway might look like:

工作表——设计渗水坑

在设计渗水坑的整体外观时，有一些事情需要考虑。如果想要一个雨水花园，但又没有足够的空间，这是一个让渗水坑看起来像雨水花园的机会。

- 如果决定在渗水坑中种植植物，可以选择不同高度、颜色和质地的耐涝的花和草（请参阅第 53 页适合种植的植物列表）。
- 像雨水花园一样，需要考虑渗水坑全年的景观效果。在整个冬天，草会保持它们的形状，但是在寒冷的月份里，花朵会消失，变得光秃秃的。
- 考虑庭院现有的景观，渗水坑与现在的花园、车道、人行道和房子搭配效果如何。
- 如果渗水坑临近道路，还要考虑渗水坑边缘线及在房屋边缘转折位置的设计。

花一分钟时间，画出理想的渗水坑大概是什么样子。

专业词汇

couple/ˈkʌpl/*n.* 对；夫妇；数个；*vi.* 结合；成婚；*vt.* 结合；连接；连合

loving/ˈlʌvɪŋ/*adj.* 亲爱的；钟情的；忠诚的

texture/ˈtekstʃə(r)/*n.* 质地；纹理；结构；本质；实质

the year round 整年地

bare/beə(r)/*adj.* 空的；赤裸的，无遮蔽的；*vt.* 露出，使赤裸

spot/spɒt/*n.* 地点；斑点；*vt.* 认出；弄脏；用灯光照射；*vi.* 沾上污渍；满是斑点

existing/ɪgˈzɪstɪŋ/*adj.* 目前的；现存的；*v.* 存在（exist 的现在分词）

against/əˈgenst/*prep.* 反对，违反；靠；倚；防备；*adj.* 不利的；对立的

current/ˈkʌrənt/*adj.* 现在的；流通的；通用的；最近的；草写的；*n.*（水、气、电）流；趋势；涌流

setback/ˈsetbæk/*n.* 挫折；退步；逆流

edge/edʒ/*n.* 边缘；优势；刀刃；锋利；*vt.* 使锐利；将…开刃；给…加上边；*vi.* 缓缓移动

rough/rʌf/*n.* 粗糙的；粗略的；粗野的；艰苦的；*vt.* 使粗糙；粗暴对待；草拟；*n.* 艰苦

Building Your Soakaway

There Are Five Main Steps to Build a Soakaway:

1. Digging your soakaway.

> **IMPORTANT**
>
> Remember, your soakaway should be located in an area that a downspout can easily be directed to (refer to page 24).
>
> ON1CALL
> CALL BEFORE YOU DIG
> It's The Law!
> www.on1call.com
> 1-800-400-2255

2. Lining your soakaway with landscape fabric.
3. Filling your soakaway with stone.
4. Creating an inlet and overflow area.
5. Finishing touches with plants and/or rocks.

> **WHEN TO SEEK PROFESSIONAL ADVICE**
>
> If your soil is high in clay (refer to page 14 for your soil analysis results), it can be more challenging to create a highly functioning soakaway. You may want to consult with a landscape professional before proceeding for advice on dealing with your specific soil conditions. Refer to Resources and look under "Soakaways" for more information.

Installing Your Soakaway

All types of soakaways are similar in design. The following describes the general installation instructions, with differences between infiltration trenches and soakaway pits noted.

Tools Required:

- Measuring tape
- Shovel
- Hammer
- Ground paint or string

Materials Required:

(Refer to page 78 for quantity calculation)

- 3/4" clear gravel
- Non-woven filter cloth or landscape fabric (enough to line the base and sides of the excavated area).
- Garden staples
- Perforated rigid plastic pipe, 10 centimetre diameter, with perforated pipe connectors (required for infiltration trenches).
- Sand-compost mix (if you have clay-heavy soil and are amending your soil.
- Decorative river rocks and/or pea gravel, sod, grasses, sedges or seedlings (your choice).

Steps:

1 Digging your soakaway

a) Once you have decided the location of your soakway project, mark the area with ground paint or string. Make sure that you are satisfied with the size and layout.

b) Dig to the appropriate depth (refer to page 70). Try to stand outside of the excavated area as much as possible when digging, and do not place any heavy machinery on the area to avoid compacting the soil.

c) If you are digging an infiltration trench, excavate the entire length of the trench at a downward slope of 3 centimetres for every 1 metre in length to help with the movement of water. To help you with this step, you may want to refer back to page 12 for instructions on calculating slope.

2 Install non-woven filter cloth or landscape fabric

Once the entire area has been dug, lay down and secure non-woven filter cloth or landscape fabric along the bottom and sides with garden staples. This prevents any soil or vegetation from entering the soakaway, while still allowing water to soak through. Be sure to leave an overhang of 50 centimetres on either side. You will use this later when completing the installation of your soakaway.

修建渗水坑

修建渗水坑有以下五个主要步骤：

1. 开挖渗水坑。

> ### 重要提示
>
> 请记住，渗水坑应该位于很容易将落水管引入的地方（请参阅第 25 页）。
>
> ON1CALL
>
> 开挖前先打个电话！
> 这是法律！
> www.on1call.com
> 1－800－400－2255

2. 渗水坑周边铺设土工织物。
3. 用石块填满渗水坑。
4. 设置引水区和溢流区。
5. 用植物或岩石装饰表面。

> ### 何时寻求专业建议
>
> 如果土壤的黏粒含量高（请参阅第 15 页的土壤分析结果），那么建造功能强大的渗水坑可能更具挑战性。在处理特殊性质的土壤之前，你可能需要咨询专业景观师。请参阅“相关资源”部分内容，查看更多“渗水坑”相关信息。

修建渗水坑

所有类型的渗水坑在设计上都是相似的。下面介绍了一般的修建说明，并注明了渗透沟和渗水坑之间的区别。

所需工具：

- 卷尺
- 铲
- 锤
- 地面油漆或细绳

所需材料：

（请参阅第 79 页，数量计算）

- 3/4 英寸（约等于 1.905 厘米）粒径的干净砾石
- 无纺布或土工织物（足够衬在开挖区域的底部和两侧）
- 花园用品
- 直径 10 厘米的穿孔硬质塑料管，穿孔硬质塑料管接头（用于渗透沟）
- 沙土混合堆肥（如果有富含黏土的土壤，要对土壤进行改良）
- 装饰性河岸岩石或小卵石、草皮、草、莎草或幼苗（根据喜好进行选择）

步骤：

❶开挖渗水坑

a）一旦确定了渗水坑的位置，用地面油漆或绳子标记该区域。确保对尺寸和布局满意。

b）开挖适当的深度（参见第 71 页）。开挖的时候尽量站在开挖区域之外，不要在该区域放置任何重型机械，以免压实土壤。

c）如果正在挖渗透沟，在整个渗透沟长度上挖出倾斜的坡度，即每 1 米长度下降 3 厘米，以帮助水流自流。为了完成这一步，请参阅第 13 页，了解坡度计算的相关说明。

❷铺设无纺布或土工织物

一旦开挖工作结束，沿着渗水坑的底部和两侧铺设无纺布或土工织物，并用专用钉固定。这样既可以防止任何土壤颗粒或植被等杂物进入渗水坑，同时还允许水流渗透。一定要在两侧留出 50 厘米的宽度，渗水坑建好后将会用到。

> **专业词汇**
>
> inlet/ˈɪnlet/*n.* 入口，进口；插入物
>
> seek/siːk/*vt.* 寻求；寻找；探索；搜索；*vi.* 寻找；探索；搜索
>
> professional/prəˈfeʃənl/*adj.* 专业的；职业的；职业性的；*n.* 专业人员；职业运动员
>
> challenge/ˈtʃælɪndʒ/*n.* 挑战；怀疑；*vt.* 向…挑战
>
> consult with 商量，协商
>
> proceed/prəˈsiːd/*vi.* 开始；继续进行；发生；行进
>
> specific/spəˈsɪfɪk/*adj.* 特殊的，特定的；明确的；详细的；[药] 具有特效的；*n.* 特性；细节；特效药
>
> measuring tape 卷尺；皮尺
>
> shovel/ˈʃʌvl/*n.* 铁铲；一铲的量；铲车；*vt.* 铲除；用铲挖；把…胡乱塞入 *vi.* 铲
>
> hammer/ˈhæmə(r)/*vi.* 锤击；敲打；重复；*n.* 铁锤；链球；音锤；*vt.* 锤击，锤打…
>
> paint/peɪnt/*vt.* 油漆；绘画；装饰；描绘；*vi.* 油漆；绘画；*n.* 油漆；颜料；绘画作品
>
> string/strɪŋ/*n.* 线，细绳；一串，一行；*vt.* 扎，缚；使排成一行，串起；伸展，拉直；*vi.* 连成一串…
>
> woven/ˈwəʊvn/*v.* 编织；交织（weave 的过去分词）编造；*adj.* 织物的；*n.* 机织织物
>
> excavate/ˈekskəveɪt/*vt.* 挖掘；开凿；*vi.* 发掘；细查
>
> excavated area　挖掘面积
>
> staple/ˈsteɪpl/*n.* 主要产品；订书针；主题；主食；*adj.* 主要的，大宗生产的；常用的；纺织纤维的…
>
> diameter/daɪˈæmɪtə(r)/*n.* 直径
>
> connector/kəˈnektə(r)/*n.* 连接器，连接头
>
> pea gravel　小卵石，豆粒砾石
>
> sod/sɒd/*n.* 草地；草皮；*vt.* 铺上草皮；以生草土覆盖；铺草皮于
>
> sedge/sedʒ/*n.* 莎草；*adj.* 莎草科的
>
> seedling/ˈsiːdlɪŋ/*n.* 秧苗，幼苗；树苗
>
> layout/ˈleɪaʊt/*n.* 布局；设计；安排；陈列
>
> machinery/məˈʃiːn(ə)ri/*n.* 机械；机器；机构；机械装置
>
> downward/ˈdaʊnwəd/*adj.* 向下的，下降的；*adv.* 向下
>
> secure/sɪˈkjʊə(r) /*adj.* 安全的；无虑的；有把握的；稳定的；*vt.* 保护；弄到；招致；缚住
>
> vegetation/ˌvedʒəˈteɪʃn/*n.* 植被；植物，草本；呆板单调的生活
>
> overhang/ˌəʊvəˈhæŋ/*vi.* 悬垂；逼近；*vt.* 悬于…之上；（危险、邪恶等）逼近；*n.*（船首或船尾）突出部分
>
> either/ˈaɪðə(r)/*adj.* 两者之中任一的；两者之中每一的；*prep.* 任何一个；*conj.* 或者

Are You Installing an **Infiltration Trench** or a **Soakaway Pit**?

3 Infiltration trench

A) Fill the bottom of the trench with 5 – 10 centimetres of 3/4" clear stone as a base, and lay the perforated pipe on top.

There are two options in this step.

B1) If you are connecting your downspouts directly to the perforated pipe you can lay the pipe on top of the stone and connect the higher end of the pipe to your roof downspout. This step may require you to attach an elbow to the base of your downspout, which you then connect to your perforated pipe (refer to downspout disconnection page 24).

B2) If you are not connecting your downspout to a perforated pipe, you can direct your downspout to the infiltration trench at ground level.

3 Soakaway pit

A) Fill 2/3 of your soakaway with 3/4" clear stone.

B1) If you are connecting your downspout(s) to your soakaway pit underground, you will have to dig a trench and connect your downspout to a solid plastic pipe that is inserted directly into the 3/4" clear stone.

B2) If you are not connecting your downspout(s) to your soakaway underground, you can direct your downspout above ground onto the soakaway.

continued on next page >

是修建一个渗透沟还是修建一个渗水坑？

❸渗透沟

A）在沟底填上5～10厘米厚的3/4英寸（约等于1.905厘米）粒径的干净砾石作为基底，然后把穿孔管放在上面。

在这个步骤中有两个选项：

B1）落水管直接连接到穿孔管上。穿孔管铺设在砾石基底层上，穿孔管的上端与屋顶的落水管连接。连接时，需要先将肘型弯头连接到落水管的底部，然后再连接到穿孔管上（请参阅第25页断开的落水管）。

B2）落水管没有连接到穿孔管上。可以把落水管直接延伸到渗透沟所在位置。

❸渗水坑

A）用3/4英寸（约等于1.905厘米）粒径的干净砾石填满渗水坑的2/3。

B1）如果要把落水管与地下的渗水坑连接起来，必须开挖一条沟渠，在落水管上连接一个硬质塑料管，塑料管末端直接插入3/4英寸（约等于1.905厘米）粒径的干净砾石里。

B2）如果没有用硬质塑料管把落水管与地下的渗水坑连接起来，则可以把落水管直接延伸到渗水坑所在位置。

专业词汇

elbow/ˈelbəʊ/*n.* 肘部；弯头；扶手；*vt.* 推挤；用手肘推开

solid/ˈsɒlɪd/*adj.* 固体的；可靠的；立体的；结实的；*n.* 固体；立方体

insert/ɪnˈsɜːt/*vt.* 插入；嵌入；*n.* 插入物；管芯；镶块；［机械］刀片

❹ Finishing your soakaway

a) Fill the remainder of the trench with 3/4" clear stone until you are 5 – 10 centimetres from the top.

b) Lay down and secure non-woven filter cloth or landscape fabric on top of the stone.

c) Either backfill with the original soil you removed in order to plant the area, or if you prefer, you can create a nice rock garden trench as shown in the picture to the right (refer to page 52 for a list of appropriate plants).

d) Install a screen on your eavestrough or a removable filter screen into your downspout in order to prevent leaf litter and twigs from entering and clogging your soakaway (refer to page 84).

How Much Stone Should I Order?
To calculate how much stone you will need to fill your soakaway, complete the calculation below.

_____metres (soakaway length) x _____ metres (soakaway width) x ______ metres (soakaway depth – should be around 1.0 metre)=

________* cubic metres of stone

CONVERSION OF CUBIC METRES TO CUBIC YARDS

Many garden centres still price their bulk aggregate (soil, sand, compost, mulch) in cubic yards. It is easy to convert your measurements:

1 cubic metre = 1.3 cubic yards

So to covert any measurements in
cubic metres to cubic yards, simply multiply by 1.3.

❹完成渗水坑

a）用 3/4 英寸（约等于 1.905 厘米）粒径的干净砾石填满沟槽的其余部分，直到距离渗水坑顶部 5～10 厘米。

b）将无纺布或土工织物铺至渗水坑顶部与所填石料齐平，并固定好。

c）如果要种植植物，可用挖出来的土回填，或者如果喜欢，也可以像右图一样修建一条漂亮的岩石花园沟渠（有关适合种植的植物列表，请参阅第 53 页）。

d）在屋檐沟槽里安装滤网，或者在落水管上安装可拆卸的滤网，以防止树叶和树枝进入并堵塞渗水坑（请参阅第 85 页）。

应该订购多少石块?

为了知道所需石块的量，需要完成下面的计算：

________ m（渗水坑长度）× ________ m（渗水坑宽度）× ________ m（渗水坑深度约 1m）= ________ m^3 的石头

立方米换算成立方码

许多园艺中心仍然以立方码为单位，对散装材料（土壤、沙子、堆肥、护根覆盖物）定价。测量值也很容易转换：

1 立方米＝1.3 立方码

因此，将所有测量值由立方米转换成立方码，只要乘以 1.3。

专业词汇

remainde/rɪ'meɪndə(r)/*n.* ［数］余数，残余；剩余物；其余的人；*adj.* 剩余的；吃剩的；*vt.* 廉价出售

backfill/'bækfɪl/*vt.* 回填；装填；*n.* 回填；用来填满坑穴的泥土或其他东西

twig/twɪg/*n.* 小枝；嫩枝；末梢；*vi.* 理解；*vt.* 理解

cubic/'kju:bɪk/*adj.* 立方体的，立方的

conversion/kən'vɜ:ʃn/*n.* 转换；变换；［金融］兑换；改变信仰

yard/jɑ:d/*n.* 院子；码（英制中丈量长度单位，1 码＝3 英尺）；庭院；帆桁；*vt.* 把…关进或围在畜栏里

bulk/bʌlk/*n.* 体积，容量；大多数，大部分，大块；*vt.* 使扩大，使形成大量；使显得重要

aggregate/'ægrɪgət/*vi.* 集合；聚集；合计；*vt.* 集合；聚集；合计；*n.* 合计；几何体；总计

convert/kən'vɜ:t/*vt.* 使转换；转换…；使…改变信仰；*vi.* 转变；变换；皈依；改变信仰

measurement/'meʒəmənt/*n.* 测量；［计量］度量；尺寸；量度制

Cost

The cost of installing a soakaway will vary depending on the size and type of soakaway.

Example: the cost of installing an infiltration trench 0.5 metres wide, 12 metres long and 1 metre deep would be:

MATERIAL	QUANTITY	COST PER UNIT	COST
3/4" clear stone	2.0 yards	$50 / yard	$100
Non-woven filtercloth / landscape fabric	1 (4ft x 100ft)	$100	$100
Landscape/Garden staples	1 (40 pack box)	$7	$7
Perforated pipe(optional)	4 (3m length)	$10	$40
Perforated pipe connectors(optional)	3	$4	$12
Downspout elbow (optional)	1	$4	$4
Flexible downspout diverters (optional)	1	$15	$15
Eavestrough filter/Screen	1	$4	$4
Monitoring/Observation well (optional)	1	$15	$15
Pea gravel or River stone (optional)	1 yard	$85	$85
TOTAL (including optional items)			**$382**
TOTAL (not including optional items)			**$211**

Example: the cost of installing a soakaway pit 1.0 metre cubed (1.0m x 1.0m x 1.0m) would be:

MATERIAL	QUANTITY	COST PER UNIT	COST
3/4" clear gravel (limestone)	1.0 yard	$50/yard	$50
Non-woven filtercloth / landscape fabric	1 (4ft x 50ft)	$45	$45
Landscape/Garden staples	1 (40 pack box)	$7	$7
Perforated pipe	1 (3m length)	$10	$10
Observation well cap on top of perforated pipe	1	$8	$8
Downspout elbow (optional)	1	$4	$4
Flexible downspout diverter (optional)	1	$15	$15
Pea gravel or River stone (optional)	¼ yard	$85	$21.25
TOTAL (including optional items)			$160.25
TOTAL (not including optional items)			$120

***NOTE: These cost tables do not include the cost of plants. If you plan on including plants into your soakaway please refer to page 52 for a list of appropriate plants. Price of plants will vary depending on what nursery or landscape center you purchase them from. As a rule of thumb, the cost of individual plants are around $5.

成本

建造渗水坑的费用根据渗水坑的大小和类型有所不同。

实例： 建造一条 0.5 米宽、12 米长、1 米深的渗透沟成本如下：

材料	数量	单位成本	总造价
3/4 英寸（约等于 1.905 厘米）粒径的干净砾石	2.0 码	50 加元/码	100 加元
无纺布/土工织物	1（4 英尺×100 英尺）（约为 121.9 厘米×3648 厘米）	100 加元	100 加元
花园专用固定钉	1（40 个每盒）	7 加元	7 加元
穿孔管（可选择）	4（3 米长）	10 加元	40 加元
穿孔管接头（可选择）	3	4 加元	12 加元
落水肘型弯头（可选择）	1	4 加元	4 加元
易弯曲的落水管转向器（可选择）	1	15 加元	15 加元
滤网	1	4 加元	4 加元
监测井/观测井（可选择）	1	15 加元	15 加元
小卵石或河石（可选择）	1 码	85 加元	85 加元
总计（包括可选择的项目）			382 加元
总计（不包括可选择的项目）			211 加元

实例： 建造一个 1 立方米的渗水坑的成本（1.0 米×1.0 米×1.0 米）如下：

材料	数量	单位成本	总成本
3/4 英寸（约等于 1.905 厘米）粒径的干净砾石（石灰岩）	1.0 码	50 加元/立方码	50 加元
无纺布/土工织物	1（4 英尺×50 英尺）（约为 121.9 厘米×1524 厘米）	45 加元	45 加元
花园专用固定钉	1（40 个每盒）	7 加元	7 加元
穿孔管（可选择）	1（3 米长）	10 加元	10 加元
多孔管顶部观测井盖	1	8 加元	8 加元
落水管肘型弯头（可选择）	1	4 加元	4 加元
易弯曲的落水管转向器（可选择）	1	15 加元	15 加元
小卵石或河石（可选择）	1/4 码	85 加元	21.25 加元
总计（包括可选择的项目）			160.25 加元
总计（不包括可选择的项目）			120 加元

*****注** 表中的成本不包括植物成本。如果打算在渗水坑中种植植物，请参阅第 53 页的植物列表。植物的价格将根据从哪个苗圃或景观中心购买而有所不同。按照经验，单株植物的成本是 5 加元左右。

专业词汇

per/强 pɜː(r)，弱 pə(r)/*prep.* 每；经；按照；每一

flexible/ˈfleksəbl/*adj.* 灵活的；柔韧的；易弯曲的

diverter/daɪˈvɜːtə/*n.* 转向器；［电］分流器；避雷针

limestone/ˈlaɪmstəʊn/*n.* ［岩］石灰岩

nursery/ˈnɜːsəri/*n.* 苗圃；托儿所；温床

as a rule of thumb 一条总要的原则，按照以往的经验，作为一个经验法则

individual/ˌɪndɪˈvɪdʒuəl/*adj.* 个人的；个别的；独特的；*n.* 个人，个体

Your Soakaway Pit Cost Table:

MATERIAL	QUANTITY	COST PER UNIT	COST
		$	$
		$	$
		$	$
		$	$
		$	$
		$	$
		$	$
		$	$
		$	$
		$	$
		$	$
TOTAL			**$**

渗水坑成本计算表：

材料	数量	单位成本	总成本
		加元	加元
		加元	加元
		加元	加元
		加元	加元
		加元	加元
		加元	加元
		加元	加元
		加元	加元
		加元	加元
		加元	加元
		加元	加元
总计			加元

Maintenance

Soakaways require very little maintenance to ensure proper operation. The only maintenance requirements include regular inspections and cleaning of inlets to prevent clogging, and inspection of monitoring wells (if installed). Sediment and leaf buildup will reduce the infiltration rate, so should be cleaned at least twice a year. Soggy ground or water pooling may be indicators that there is a clog. Filters on the eavestroughs or at the inlet of the trench are great in preventing leaf litter from entering (refer to the pictures to the right). These can be purchased at your local hardware or landscaping supply store.

If you have planted your soakaway to look like a garden, refer to page 60 for tips on garden maintenance.

维护

渗水坑在日常使用中很少需要维护。唯一的维护工作就是定期检查和清理入水口，以防止堵塞，以及检查监测井（如果有的话）。沉积物和树叶堆积会降低渗水坑的渗透率，所以应该每年至少清理两次。潮湿的地面或积水可能是堵塞的迹象。沟槽两侧和沟槽入水口处的过滤器可以很好地防止落叶进入（见右图）。这些零件可以在当地的五金店或景观用品商店购买。

如果渗水坑已种植得像花园一样，请参阅第 61 页关于花园维护的建议。

专业词汇

regular/'regjələ(r)/*adj.* 定期的；有规律的；合格的；整齐的；*n.* 常客；正式队员；中坚分子

inspection/ɪn'spekʃn/*n.* 视察；检查；安全检测

soggy/'sɒgi/*adj.* 浸水的；透湿的；沉闷的

indicator/'ɪndɪkeɪtə(r)/*n.* 指示器；［试剂］指示剂；［计］指示符；压力计

hardware/'hɑːdweə(r)/*n.* 计算机硬件；五金器具

Permeable Pavement

In a typical residential neighbourhood, 50% or more of the surfaces are impermeable. These surfaces include driveways, sidewalks, roofs, patios and decks. Water can't pass through these surfaces, but rather, flows over them. This means that over 50% of the rain water that lands on our property is potentially running over an impermeable surface and into a storm drain.

By installing permeable pavement, we can have a huge impact on reducing stormwater runoff and keeping the water that falls on our property, on our property. In fact, permeable pavement can reduce the amount of runoff by 30% to 100%, when compared to conventional pavement.

Permeable pavement is an attractive alternative hard surface feature. When used on a driveway it can greatly reduce the amount of impermeable area on a property.

Permeable pavement allows water to drain through it, rather than running over it. A stone reservoir beneath the surface collects the water and allows it to slowly soak into the surrounding soil.

Why Is Permeable Pavement a Good Option?

Permeable pavement provides many important environmental benefits. These include:

- **Reducing water pollution**
  By reducing runoff, pollutants found on the surface of the pavement are not swept into the storm drains and into local water courses. They are trapped in the pores of the permeable pavement or the underlying stone reservoir.

- **Keeping rain water on our property**
  The amount of water entering our water courses during storm events is reduced because the water is being kept where it lands, rather than running off into storm drains causing high water levels and stream bank erosion.

- **Reduced winter maintenance**
  The design of permeable pavement greatly reduces snow and ice accumulation on its surface. This is due to the fact that they are freely draining structures that don't pond snowmelt between freeze-thaw cycles like regular asphalt driveways do, therefore less ice buildup occurs.

IS PERMEABLE PAVEMENT RIGHT FOR ME?

	YES	NO
Are you planning to install or upgrade a driveway, walkway or patio?	☐	☐
Is the slope of the area where you plan to install the permeable pavement more than 1% and less than 5% (refer to page 12 for instructions on calculating slope)?	☐	☐

If you answered 'YES' to both questions, you may be a candidate for permeable pavement.

PERMEABLE PAVEMENT NOT FOR YOU?

Permeable pavement can be expensive and often requires a contractor to do the work. However, you can still do a few things on your property to reduce the number of impermeable surfaces. Consider...

- Removing concrete pathways and patios and replacing them with permeable pavers. Allow grass or moss to grow up between them, giving water a place to be absorbed.
- Add grass to your driveway. Why not remove part of the paved surface and, instead, just have two paved areas for tires?
- Change paved surfaces to gravel. Gravel slows down moving water, giving it more opportunity to be absorbed into the ground.

透水铺装

在一个典型的居民区，50％或更多的表面是不透水的，包括行车道、人行道、屋顶、天井和露天平台。雨水在这些表面上不能渗透，只能形成径流。这意味着，降落在庭院中的50％以上的雨水都通过不透水地面形成径流，排入雨水管道。

铺设透水路面，对减少雨水径流和滞蓄雨水都会产生巨大的影响。事实上，与传统路面相比，透水铺装可以减少 30％～100％的雨水径流。

透水铺装是一种优良的硬质铺装替代方案。应用到车道上时，可以大大减少庭院中不透水区域的面积。

透水铺装允许雨水通过它，而不是从表面流过。透水铺装所形成的地下蓄水池能够收集雨水，并使雨水慢慢渗入周围的土壤中。

为什么透水铺装是好的选择？

透水铺装具有显著的环境效益，包括：

- **减少水污染**

通过减少径流量，道路表面的污染物就不会被冲进雨水管道，也不会进入当地河道。污染物通过透水铺装或地下蓄水池的孔隙进行过滤或净化。

- **滞留雨水**

在暴雨期间，雨水被就地滞留，渗入地下，进入河道的水量就会减少，不会造成河流水位抬升或河岸侵蚀。

- **减少冬季维护**

透水铺装设计有效地减少了路面的冰雪堆积。这是由于透水铺装的自由排水结构设计，不会像普通的沥青车道那样发生冻融循环而造成路面破坏。

透水铺装适合我吗？	是	否
你是否打算修建或改良行车道、人行道或露台？	□	□
你计划建设透水铺装的区域坡度是否大于1％小于 5％（请参阅第 13 页有关“坡度计算”的说明）？	□	□

如果对这两个问题都回答“是”，那么透水路面适用于你的庭院。

透水铺装不适合你？

透水铺装可能很昂贵，而且通常需要专门的承包商来施工。然而，你仍然可以在庭院中做一些事情来减少不透水路面的面积。仔细考虑一下……

- 可以移除混凝土路面和露台，用渗透性好的铺路材料替换。草或苔藓可在其中生长，雨水也有了入渗的地方。
- 在行车道上种植草坪。为什么不移除部分铺好的路面，而只留下两条窄长的路面供汽车轮胎通过？
- 将铺好的路面改为砾石路面。砂砾减慢了水流的速度，使雨水有更多的机会渗入地下。

专业词汇

neighbourhood/ˈneɪbəhʊd/*n.* 临近；周围；临近关系；附近一带

roof/ruːf/*n.* 屋顶；最高处，顶部；最高限度

patio/ˈpætiəʊ/*n.* 露台；天井

deck/dek/*n.* 甲板；行李舱；露天平台；*vt.* 装饰；装甲板；打扮

land/lænd/*vi.* 登陆；到达；*vt.* 使…登陆；将…卸下；使…陷于；*n.* 国土；陆地；地面

runoff/ˈrʌnɒf/*n.* ［水文］径流；决赛；流走的东西；*adj.* 决胜的

conventional/kənˈvenʃənl/*adj.* 符合习俗的；传统的；常见的；惯例的

sweep/swiːp/*vt.* 扫除；猛拉；掸去；*vi.* 扫，打扫；席卷；袭击；*n.* 打扫，扫除；范围

swept/swept/*v.* 扫除（sweep 的过去式和过去分词）；*adj.* 扫频的

water course ［水文］水道；［水文］河道

trap/træp/*vt.* 诱捕；使…受限制；使…陷入困境；*vi.* 设陷阱…；*n.* 陷阱；圈套［建］存水弯

pore/pɔː(r)/*n.* 气孔；小孔；*vi.* 细想；凝视；熟读；*vt.* 使注视

underlying/ʌndəˈlaɪɪŋ/*adj.* 潜在的；根本的；在下面的；优先的；*v.* 放在…的下面；为…的基础；优先于

stream/striːm/*n.* 溪流；流动；潮流；光线；*vi.* 流；涌进；飘扬；*vt.* 流出；涌出；使飘动

bank/bæŋk/*n.* 银行；岸；浅滩；储存；*vt.* 将…存入银行；倾斜转弯；*vi.* 堆积；倾斜转弯

accumulation/əˌkjuːmjəˈleɪʃn/*n.* 积聚，累积；堆积物

pond/pɒnd/*vt.* 筑成池塘；*vi.* 筑成池塘；*n.* 池塘

snowmelt/ˈsnəʊmelt/*n.* ［水文］融雪水

freeze/friːz/*n.* 冻结；凝固；*vi.* 冻结；冷冻；僵硬；*vt.* 使…冻住；使…结冰

thaw/θɔː/*n.* 解冻；融雪；*vi.* 溶解；变暖和；*vt.* 使溶解；使变得不拘束

asphalt/ˈæsfælt/*n.* 沥青；柏油；*vt.* 以沥青铺；*adj.* 用柏油铺成的

buildup/ˈbɪldʌp/*n.* 增强；发展；形成；组合

upgrade/ˈʌpgreɪd/*vt.* 使升级；提升；改良品种；*n.* 升级；上升；上坡；*adv.* 往上 *adj.* 向上

candidate/ˈkændɪdeɪt/*n.* 候选人，候补者；应试者

contractor/kənˈtræktə(r)/*n.* 承包人；立契约者

concrete/ˈkɒŋkriːt/*adj.* 混凝土的；实在的；具体的；有形的；*vi.* 凝结；*vt.* 使凝固；用混凝土修筑

pathway/ˈpɑːθweɪ/*n.* 路，道；途径，路径

paver/ˈpeɪvə(r)/*n.* 铺砌工；铺路材料；铺路（料）机

moss/mɒs/*n.* 苔藓；泥沼；*vt.* 使长满苔藓

Getting Started

There are three main types of permeable pavement:

1. **Permeable Pavers**
Similar to interlocking stone pavers, permeable pavers are designed with special gaps between them. These gaps are filled with small stones that allow the water to percolate through.

Advantages:
- Easy for contractors to install.
- Installation is not weather dependent.
- Very high water infiltration rates.
- Easy to maintain and repair.
- Can be plowed in the winter.

Disadvantages:
- Higher cost.

2. **Plastic or Concrete Grid Systems**
Durable plastic or concrete grids are filled with gravel or a permeable soil mix for low growing plant material (like grass or moss). The grid supports vehicle traffic and prevents compaction of the fill material.

Advantages:
- Plastic grid systems usually made from 100% recycled plastic.
- Easy to install.
- Doesn't require an even surface.

Disadvantages:
- Does require some maintenance (needs to be mowed).
- Grass pavers require substantial watering in the first year.
- Snow removal may require specialized equipment (i.e. raised plow edge and rollers).

3. **Porous Asphalt**
Porous asphalt is specially designed with spaces that allow water to drain through it.

Advantages:
- Lower cost.
- Uses same equipment as laying conventional asphalt.

Disadvantages:
- Installation is weather dependent.
- Susceptible to water damage.
- Lower relative strength than other applications.

Installing Permeable Pavement

Installing permeable pavement is not a do-it-yourself project. You will be required to hire a professional. However, if you are planning to install or upgrade a driveway, walkway or patio anyway, why not consider an option that is also good for the environment.

Here are some considerations to be aware of before you proceed:

Underground Utilities
Call your local utility company before you dig. This is especially the case before installing permeable pavement because it has a deeper base than conventional pavement so shallow utilities may be impacted.

IMPORTANT
CALL BEFORE YOU DIG!

ON1CALL
CALL BEFORE YOU DIG
It's The Law!
www.on1call.com
1-800-400-2255

Setbacks from Buildings
Permeable pavement can only be installed in areas that are sloped downhill from building foundations, with a slope of less than 5%. In most cases, you will not need to include a setback from your home's foundation. However, if the surface receives rainwater runoff from other surfaces, like a disconnected downspout, you will have to leave a setback of 4 metres from your home to the start of your permeable pavement installation.

Soil Type
Permeable pavement systems can be installed on any soil type. However,

开始

透水铺装主要有三种类型：

1. 透水砖

类似于互相锁扣的石质铺路材料，透水砖之间设计有特殊的缝隙。这些缝隙中充满了小石头，可以让雨水渗透进去。

优点：

- 便于承包商施工。
- 天气条件对施工条件影响小。
- 水的渗透速率非常高。
- 易于维护和维修。
- 可以在冬天施工。

缺点：

- 成本比较高。

2. 塑料或混凝土格栅植草砖

耐用的塑料或混凝土格栅中填充了砾石等渗透性土壤混合物，便于植物生长（如草或苔藓）。该格栅植草砖支持车辆通行，也可以防止填充的材料被压实。

优点：

- 塑料格栅通常由百分之百可回收塑料制成。
- 施工简便。
- 不需要平整地面。

缺点：

- 需要一些维护（需要割草）。
- 第一年草坪需要大量浇水。
- 除雪可能需要专门的设备（例如安装在滚轮上凸起的犁刃）。

3. 透水沥青

多孔沥青设计有足够的空隙让雨水渗透。

优点：

- 成本较低。
- 使用与铺设常规沥青路面相同的设备。

缺点：

- 天气条件对施工条件影响大。
- 容易受水浸泡而产生损害。
- 强度低于其他类型的透水路面。

透水铺装

透水铺装不是一项能自己完成的工程，需要雇佣专业人员来帮忙。如果你打算修建或升级改造行车道、人行道或露台，为什么不考虑一个更有利于环境的选择呢？

下面是你开工之前需要注意的事项：

地下设施

开挖之前，给当地的公用事业公司打一个电话，透水铺装工程施工前更要如此。因为透水铺装比传统路面铺装的基础更深，浅层的公共设施可能会受到影响。

重要提示

开挖前先打个电话！

ON1CALL

开挖前先打个电话！
这是法律！
www. on1call. com
1－800－400－2255

透水铺装的适用条件

透水铺装只适用于建筑地基向外倾斜坡度小于5%的地方。大多数情况下，房屋都能满足要求。如果透水路面还要接纳来自其他汇水面的雨水径流，比如一根断开的落水管中的雨水，那么透水路面需要距离房屋至少4米。

土壤类型

透水路面可以铺设在任何类型的土壤上。然而，

专业词汇

interlock/ˌɪntəˈlɒk/*v.* ［计］互锁；连锁

interlocking *adj.* 连锁的；*n.* 连锁；咬合作用

percolate/ˈpɜːkəleɪt/*vi.* 过滤；渗出；浸透；*vt.* 使渗出；使过滤；*n.* 滤过液；渗出液

plow/plaʊ/*vi.* ［农机］犁；耕地；破浪前进；开路；*vt.* ［农机］犁；耕；开路；

grid/grɪd/*n.* 网格；格子，栅格；输电网

durable/ˈdjʊərəbl/*adj.* 耐用的，持久的；*n.* 耐用品

even/ˈiːvn/*adj.* ［数］偶数的；平坦的；相等的；*adv.* 甚至；即使；还；实际上；*vt.* 使…

mow/məʊ/*n.* 草堆；皱眉；谷堆；*vt.* 割草；收割庄稼；*vi.* 割草；收割庄稼

mowed *v.* 用镰刀等刈草；割草（mow 的过去分词）

substantial/səbˈstænʃl/*adj.* 大量的；实质的；内容充实的；*n.* 本质；重要材料

removal/rɪˈmuːvl/*n.* 免职；移动；排出；搬迁

raised/reɪzd/*adj.* 凸起的；发酵的；浮雕的；*v.* 提高（raise 的过去时和过去分词）

roller/ˈrəʊlə(r)/*n.* ［机］滚筒；［机］滚轴；棍子；滚转机

porous/ˈpɔːrəs/*adj.* 多孔渗水的；能渗透的；有气孔的

susceptible/səˈseptəbl/*adj.* 易受影响的；易感动的；容许…的

susceptible to *adj.* 易受…影响的；对…敏感的

relative/ˈrelətɪv/*adj.* 相对的；有关系的；成比例的；*n.* 亲戚；相关物；［语］关系词；亲缘植物

hire/ˈhaɪə(r)/*vt.* 雇用；出租；*vi.* 雇用，租用；受雇；*n.* 雇用；租用；租金；工钱

consideration/kənˌsɪdəˈreɪʃn/*n.* 考虑；原因；关心；报酬

considerations *n.* 考虑，注意事项（consideration 的复数形式）；体贴

shallow/ˈʃæləʊ/*adj.* 浅的；肤浅的；*n.* ［地理］浅滩；*vt.* 使变浅；*vi.* 变浅

if your soil type is high in clay (refer to page 14 for instructions on soil assessment) an underdrain is recommended. An underdrain is a perforated pipe located under the pavement that helps to remove excess water from the system.

There are Five Main Layers in a Permeable Pavement System.

Layer 1 –Soil Base
Your contractor will dig down roughly 0.5 metres to begin to install your permeable pavement system. The base should be completely flat, but care should be taken not to compact the soil base, as this will reduce its ability to absorb water. Any heavy machinery used for excavation should be positioned outside of the excavated area.

Your contractor may install a geotextile (permeable landscape fabric) along the bottom and sides of the base. This will help to prevent clogging and maintain

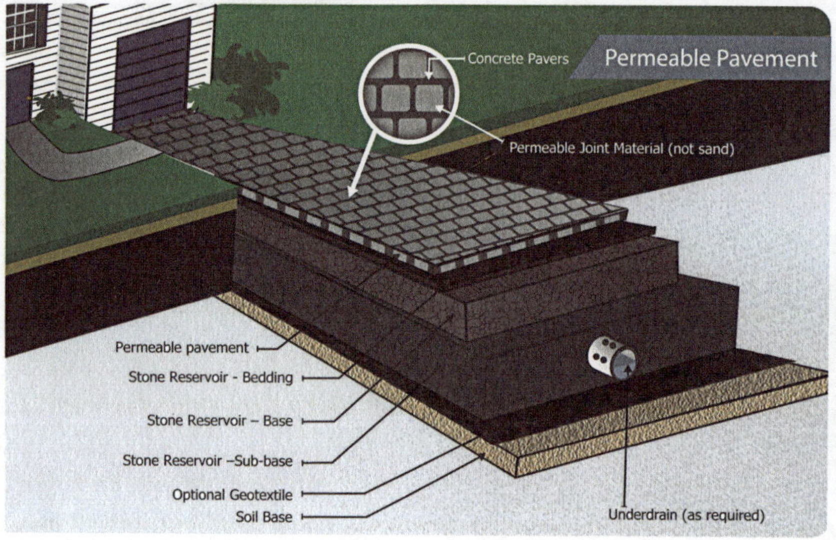

water flow between the soils and the stone reservoir.

Layer 2 – Stone Reservoir –Sub-base
The sub-base layer consists of medium-sized crushed stone (approximately 50 millimetres in diameter). If you are on clay soils and require an underdrain, it will be installed within this layer.

Layer 3 – Stone Reservoir – Base
The base layer consists of small crushed stone (approximately 20 millimetres in diameter).

Layer 4 – Stone Reservoir - Bedding
The bedding layer is typically between 4 and 7.5 centimetres in depth. It consists of tiny rock pieces, approximately 0.5 centimetres in diameter.

Layer 5 – Permeable Pavement
The final layer of a permeable pavement system is the surface layer. This is where your contractor will install your choice of permeable pavers, plastic/concrete grid system or porous asphalt.

CLEAN FILL REQUIRED

The crushed stone used in the construction of permeable absorbent pavement systems isn't just regular stone – it is a specially formulated clean-fill mix that contains no small sediment pieces. This ensures that the spaces between the stones remain open for water to travel through, rather than getting clogged with dirt and sediment.

How Much Will It Cost?

There are many site specific factors to consider when costing out a permeable pavement system. This is a general comparison of permeable pavement types with cost estimates based on 2011 data.

Plastic/Concrete Grid Systems:
Range in price from approximately $4.95 to $18.90 per metre squared of installed pavement.

Pervious Concrete/Porous Asphalt:
Can cost from $6.60 to $21.45 per metre squared of installed pavement.

Permeable Pavers:
Range in price from $16.50 to about $33.00 per metre squared of installed pavement.

In general, the cost of a permeable pavement system is 2 to 3 times more than installing conventional asphalt.

(Paver Search Inc., 2011)

如果土壤黏粒含量很高（请参阅第 15 页的土壤评估说明），建议使用地下排水系统。地下排水系统是位于路面下的穿孔管道，用来帮助清除地基中多余的水分。

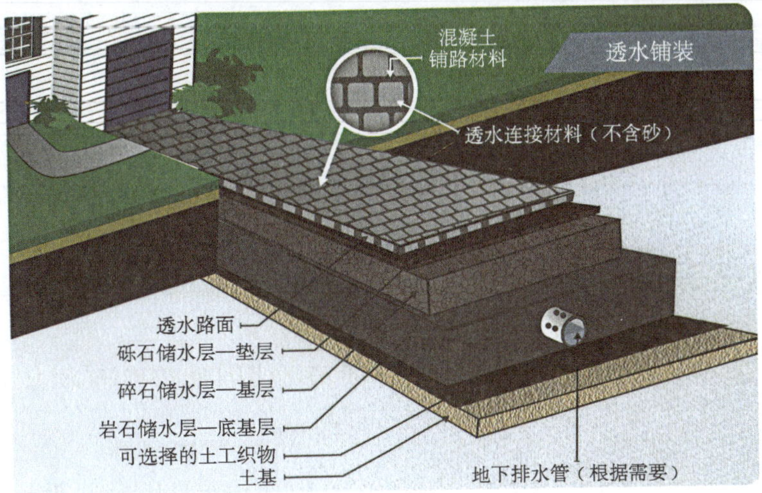

透水铺装系统主要分为 5 层，如右图所示。

第一层——土基

工程承包商会挖深大约 0.5 米，然后开始透水铺装系统施工。地基土应非常平整，但注意不要压实，因为这会降低它的透水能力。任何用于挖掘的重型机械都应放置在挖掘区域之外。

承包商会在土基的底部和侧面铺设土工织物（可渗透的无纺布），这将有助于防止阻塞，并保持水在土壤和蓄水池之间流动。

第二层——岩石储水层-底基层

底基层由中等大小的碎石（直径约 50 毫米）组成。如果底基层在黏性土层上，则需要在该层设置一个地下排水管。

第三层——碎石储水层-基层

基层由小碎石（直径约 20 毫米）组成。

第四层——砾石储水层-垫层

垫层厚度一般厚 4～7.5 厘米，由直径约为 0.5 厘米的小石块组成。

第五层——透水路面

透水铺装最上面的一层即表层，承包商将铺设你选择的透水砖、塑料或混凝土格栅植草砖或多孔沥青。

填充材料

被用于透水路面的碎石不是普通的石头，而是一种特殊配方的清洁混合材料，不包含小的颗粒物。这就确保了石头之间的空隙持续连通，水可以顺利下渗，而不会被污垢和沉积物堵塞。

要花多少钱？

在计算透水铺装成本时，需要考虑许多具体的场地因素。基于 2011 年的数据估算的透水铺装成本与普通路面铺装成本的比较如下：

塑料或混凝土格栅植草砖：

每平方米的价格为 4.95～18.90 加元。

透水混凝土/多孔沥青：

每平方米的成本从 6.60 加元到 21.45 加元不等。

透水砖：

每平方米的成本从 16.50 加元到 33.00 加元不等。

一般来说，透水铺装的成本是传统沥青路面的 2～3 倍。

（相关价格信息由铺路材料公司提供，2011 年）

专业词汇

underdrain/ˌʌndəˈdreɪn/*n*. 暗渠；阴沟；*vt*. 开阴沟排水

roughly/ˈrʌfli/*adv*. 粗糙地；概略地

sub-base *n*. 底基层；垫板

medium/ˈmiːdiəm/*adj*. 中间的，中等的；半生熟的；*n*. 方法；媒体；媒介；中间物

crushed stone ［建］碎石

bedding layer 垫层

formulate/ˈfɔːmjuleɪt/*vt*. 规划；用公式表示；明确地表达

formulated 按配方制造

tiny/ˈtaɪni/*adj*. 微小的；很少的

dirt/dɜːt/*n*. 污垢，泥土；灰尘，尘土；下流话

pervious/ˈpɜːvɪəs/*adj*. 能被通过的；能接受的；可渗透的

WORKSHEET - Getting Quotes

It is recommended that you get three quotes for any large project you undertake. Ask for references of other residential projects that you can contact as well.

Quote 1

Name of Company: ______________________

Phone #: ______________________

Email Address: ______________________

Contact Name: ______________________

Project estimate (quote): $ ______________________

Reference 1: ______________________

Name: ______________________

Phone #: ______________________

Reference 2: ______________________

Name: ______________________

Phone #: ______________________

Reference 3: ______________________

Name: ______________________

Phone #: ______________________

Notes: ______________________

Quote 2

Name of Company: ______________________

Phone #: ______________________

Email Address: ______________________

Contact Name: ______________________

Project estimate (quote): $ ______________________

Reference 1: ______________________

Name: ______________________

Phone #: ______________________

Reference 2: ______________________

Name: ______________________

Phone #: ______________________

Reference 3: ______________________

Name: ______________________

Phone #: ______________________

Notes: ______________________

Quote 3

Name of Company: ______________________

Phone #: ______________________

Email Address: ______________________

Contact Name: ______________________

Project estimate (quote): $ ______________________

Reference 1: ______________________

Name: ______________________

Phone #: ______________________

Reference 2: ______________________

Name: ______________________

Phone #: ______________________

Reference 3: ______________________

Name: ______________________

Phone #: ______________________

Notes: ______________________

Calculating the Cost of Your Project

Project cost = Total metres squared of your project area X Cost of the permeable pavement (as per your quotes)

Project cost = __________ m^2 X $__________ per m^2 **Project cost** = $ ______________________

工作表——报价

建议进行任何大型项目时，都要获取三份报价。也可以尽可能联系其他同类项目作为参考资料。

报价 1

公司名称：____________________

电话号码：____________________

电子邮件地址：____________________

联系人姓名：____________________

项目估价（报价）：加元____________________

参考 1：____________________

名称：____________________

电话号码：____________________

参考 2：____________________

名称：____________________

电话号码：____________________

参考 3：____________________

名称：____________________

电话号码：____________________

注意事项：____________________

____

____

报价 2

公司名称：____________________

电话号码：____________________

电子邮件地址：____________________

联系人姓名：____________________

项目估价（报价）：加元____________________

参考 1：____________________

名称：____________________

电话号码：____________________

参考 2：____________________

名称：____________________

电话号码：____________________

参考 3：____________________

名称：____________________

电话号码：____________________

注意事项：____________________

____

____

报价 3

公司名称：____________________

电话号码：____________________

电子邮件地址：____________________

联系人姓名：____________________

项目估价（报价）：加元____________________

参考 1：____________________

名称：____________________

电话号码：____________________

参考 2：____________________

名称：____________________

电话号码：____________________

参考 3：____________________

名称：____________________

电话号码：____________________

注意事项：____________________

____

____

计算项目成本

项目成本＝项目面积（平方米）×透水铺装成本单价（根据你的报价）

项目成本＝____________平方米 × $____________每平方米　**项目成本**＝$____________

专业词汇

getting/ˈgetɪŋ/*v.* 获得；了解（get 的现在分词）；抓住；*n.* 获得；采煤，采掘

quote/kwəʊt/*vt.* 报价；引述；举证；*vi.* 报价；引用；印证；*n.* 引用

estimate/ˈestɪmət/*vi.* 估计，估价；*vt.* 估计，估量；判断，评价；*n.* 估计，估价；判断，看法

Maintenance

No matter which type of permeable pavement you choose, maintaining it can be a relatively easy process, providing that you do it regularly. Well maintained permeable pavement should last at least 20 years.

How to maintain your permeable pavement

Don't Spread Anti-slip material such as sand and gravel

Anti-slip material, such as gravel, sand, or cat litter, typically used on driveways to improve winter traction, will clog the spaces in the permeable pavement that allow the water to pass through. Ice does not tend to form on permeable pavement, so these products are not required.

Use salt sparingly

Permeable pavement requires 75% less salt than conventional pavement. If you do need to use salt, the environmentally friendly brands are recommended.

Get out the vacuum

If you notice an area of your permeable pavement is getting clogged and the water is no longer moving through it (usually where the tires of your car sit), vacuum it with a shop vac. This will remove any sediment build up in the pores and allow water to move through it again. If this doesn't work, use a pressure washer and some biodegradable soap.

Weeding

Should weeds start to grow in the spaces between the pavers, pull them out. They can reduce the rate of infiltration through the pavement.

Get out the lawn mower

If you chose to use the plastic or concrete grid system, you will need to mow it, just like your regular lawn. The grass will also require watering, especially in the first year as it is getting established.

Shovel or plow after a snow storm

Permeable pavement can be shoveled or plowed like any other pavement. However, it is suggested that you raise your plow blade about 5 centimetres higher than normal.

Do not pile snow on your permeable pavement after shoveling. This could lead to clogging when the snow melts in the spring.

No seal coating required

You should never apply a seal coating to your permeable pavement.

HOLD THE SALT

Compared to asphalt driveways, permeable pavement actually reduces the number of pollutants getting into the groundwater. However, permeable pavement does allow salt that is applied to its surface to migrate more easily into the underlying soils. These salts can increase the mobility of heavy metals (e.g. lead, copper, cadmium). As a result, the salinity and heavy metal content moving into the groundwater can increase if the groundwater is too close to the base of the pavement structure.

For this reason road salt is not recommended for permeable pavement systems. On a well maintained permeable pavement structure, little to no salt should be required in the colder months. The design of permeable pavement actually reduces the occurrence of black ice and frozen puddles.

维护

无论选择哪种类型的透水铺装，都需要定期维护，维护工作比较容易。维护良好的透水路面应至少可以使用 20 年。

如何维护透水铺装

不要铺防滑材料，如砂子和砾石

防滑材料，如砾石、砂子或猫砂，通常用在行车道上以改善冬季车辆的牵引力。这些防滑材料容易堵塞透水路面的空隙。透水路面不容易结冰，所以不需要使用这些防滑材料。

尽量少用盐

透水路面比传统路面可以少用 75%的盐。如果确实需要使用盐，推荐使用环保品牌。

使用真空吸尘器

如果发现透水路面有一个地方被堵塞了，导致雨水不能通过（通常是汽车轮胎经过的地方），用真空吸尘器吸一下。这将去除孔隙里积聚的沉淀物，让水再次顺利流过。如果这个方法不奏效，可以使用压力垫圈和一些可生物降解的肥皂。

除草

如果透水砖之间开始长杂草，就把它们拔掉，因为它们会降低路面的渗透速度。

使用割草机

如果选择使用塑料或混凝土格栅植草砖，就像平常的草坪一样需要修剪，也需要浇水，尤其是刚长出来的第一年。

暴雪后使用除雪铲或犁

透水路面可以像其他路面一样使用除雪铲或犁。建议把犁刀抬高大约 5 厘米。

铲雪后，不要把雪堆在透水路面上，因为当春天雪融化时，会导致透水路面堵塞。

无需密封涂层

不要在透水路面上再涂一层密封涂层。

留住盐分

与沥青车道相比，透水路面实际上减少了进入地下水的污染物数量。但是，透水路面确实让盐更容易从表面迁移到深层土壤中。这些盐可以增加重金属（如铅、铜、镉）的流动性。因此，如果地下水水位非常接近路面结构的基础，则进入地下水中的盐分和重金属含量就会增加。

因此，不建议在透水路面使用道路盐。在维护良好的透水路面结构上，在较冷的月份应少用或不用盐。透水路面的设计实际上减少了道路黑冰和冰冻坑的出现。

专业词汇

spread/spred/*vi.* 传播；伸展；*vt.* 传播，散布；展开；伸展；铺开；*n.* 传播；伸展

anti/ˈænti/*adj.* 反对的；*n.* 反对者，反对论者

slip/slɪp/*vi.* 滑动；滑倒；犯错；失足；*vt.* 使滑动；滑过；摆脱；塞入；*n.* 滑，滑倒

anti-slip *adj.* 防打滑；防滑动；防空转

cat litter　猫砂

traction/ˈtrækʃn/*n.* 牵引；[机][车辆] 牵引力

sparing/ˈspeərɪŋ/*adj.* 节约的；贫乏的；保守的；*n.* 抽出；宽恕；免去；给予

sparingly *adv.* 节俭地；保守地；爱惜地

brand/brænd/*n.* 商标，牌子；烙印；*vt.* 铭刻于，铭记；到烙印于；印…商标于

vacuum/ˈvækjuəm/*n.* 真空；空间；真空吸尘器；*adj.* 真空的；利用真空的；产生真空的；*vt.* 用真空吸

washer/ˈwɒʃə(r)/*n.* [机] 垫圈；洗涤器；洗衣人

biodegradable/ˌbaɪəudɪˈgreɪdəbl/*adj.* 生物所能分解的，能进行生物降解的

weed/wiːd/*vt.* 除草；铲除；*n.* 杂草，野草；菸草；*vi.* 除草

pull/pul/*vt.* 拉；拔；拖；*vi.* 拉；拖；拔；*n.* 拉，拉绳；拉力，牵引力

pull out　离开，撤离；拔出；度过难关；恢复健康

lawn　mower　割草机

establish/ɪˈstæblɪʃ/*vt.* 建立；创办；安置；*vi.* 植物定植

seal/ˈsiːl/*n.* 密封；印章；海豹；封条；*vt.* 密封；盖章；*vi.* 猎海豹

coating/ˈkəutɪŋ/*n.* 涂层；包衣；衣料；*v.* 给…穿上外套；盖上（coat 的 ing 形式）

migrate/maɪˈgreɪt/*vi.* 移动；随季节而移居；移往；*vt.* 使移居；使移植

mobility/məuˈbɪləti/*n.* 移动性；机动性；[电子] 迁移率

lead/liːd/*n.* 领导；铅；导线；榜样；*vt.* 领导；致使；引导；指挥；*vi.* 领导；导致

copper/ˈkɒpə(r)/*n.* 铜；铜币；警察；*adj.* 铜的；*vt.* 镀铜于

cadmium/ˈkædmiəm/*n.* [化学] 镉（元素符号 Cd）

salinity/səˈlɪnəti/*n.* 盐度；盐分；盐性

content/kənˈtent/*n.* 内容，目录；满足；容量；*adj.* 满意的；*vt.* 使满足

black ice　[水文] 黑冰（一种硬而滑的透明薄冰）

puddle/ˈpʌdl/*n.* 水坑，泥潭；胶土（由黏土与水和成，不透水）；*vt.* 使泥泞，搅浑；把…捣成胶土；搅炼…

Connecting the Drops

All of the projects in the guide can be used as building blocks. You can connect them with each other to create dynamic systems that allow you to store and use rain water on your property. Here is just a sample of some of the connections that are possible.

Disconnect Downspout → Infiltration Trench → Soakaway Pit

Photo Credit: Sean James, Fern Ridge Landscaping

Disconnected Downspout → Infiltration Trench

雨水的连通路径

本手册中所有的项目都可以用作构建模块。你可以把它们联系起来创造出动态系统，在住宅和庭院中储存和使用雨水。下面是一些连通路径的示例。

蕨类景观　　图片来源：肖恩·詹姆斯

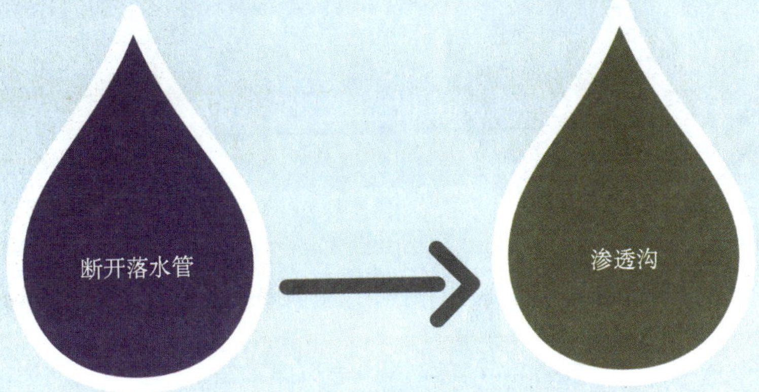

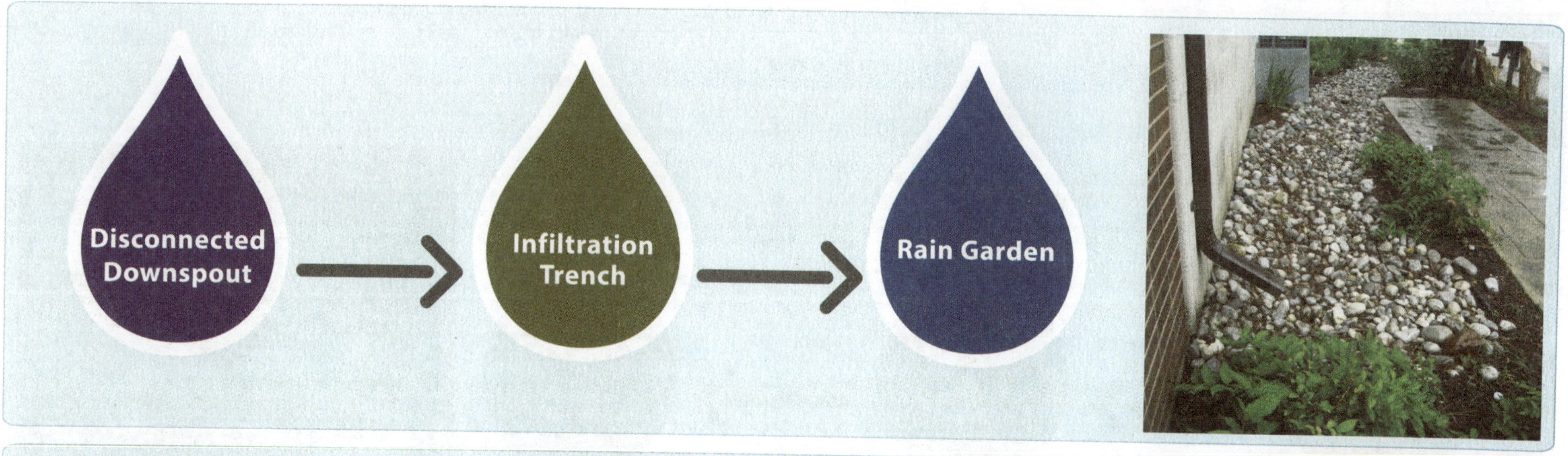
Disconnected Downspout
Infiltration Trench
Rain Garden

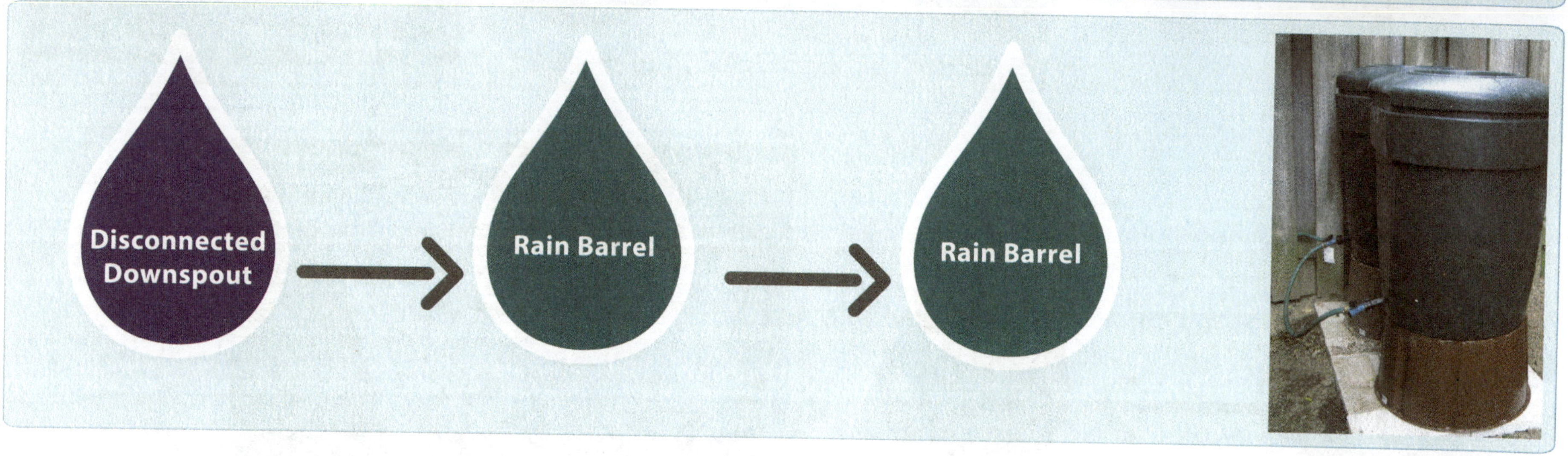
Disconnected Downspout
Rain Barrel
Rain Barrel

断开落水管
渗透沟
雨水花园

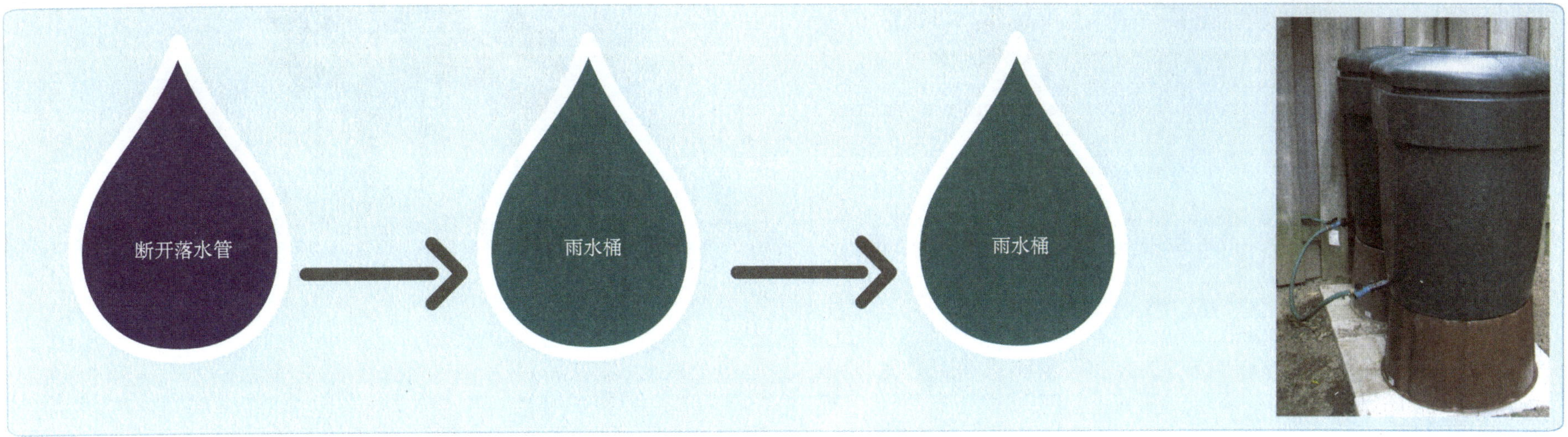
断开落水管
雨水桶
雨水桶

Connecting the Drops on your own Property

Now that you have seen some of the projects that are possible, turn back to the property sketch you created on page 8 and sketch in your ideas.

Some things you will want to consider when choosing your projects and how to connect them are:

- How much space do you have?
- How big of a project do you want to undertake? (Are you looking to do a major garden makeover this spring, or perhaps just a simple weekend project)
- How much maintenance work can you commit to doing?
- How creative do you want to get?

Remember, your only limitation is your imagination. Enjoy!

在你的庭院中将雨水连通起来

现在你已经看到了一些可行的项目，回到第 9 页所创建的场地草图，并将你的想法画在上面。

在选择项目时需要考虑以下事项并将它们联系起来：

- 你有多大的空间？
- 你想要承担多大的项目（你是想在今年春天做一个大的花园改造项目，还是仅仅想做一个简单的周末项目）？
- 你能做多少维护工作？
- 你有多大的创造力？

请记住，你唯一的限制就是你的想象力，享受这个过程吧！

Resources

Naturalization

Ontario Invasive Plants Council – Grow Me Instead
www.ontarioinvasiveplants.ca

Region of Peel – Fusion Landscaping
www.peelregion.ca/watersmartpeel

City of Toronto – Natural Lawn and Garden Care
www.toronto.ca/pesticides

York Region – Water for Tomorrow
www.york.ca/waterfortomorrow

TRCA – Healthy Yards Resources
www.trca.on.ca/yards

Canadian Wildlife Federation – Wild about Gardening
www.wildaboutgardening.org

Evergreen – Native Plant Database
www.nativeplants.evergreen.ca

Where to Buy Native Plants

It is recommended that native plants are used whenever appropriate. Native plants can be found at garden centres, nurseries and at local native plant sales.

North American Native Plant Society – Commercial Native Plant Growers
www.nanps.org

Ontario Invasive Plant Council – Nursery Recognition Program
www.ontarioinvasiveplants.ca

相关资源

移植部分相关资源

安大略入侵植物委员会——替代植物
www.ontarioinvasiveplants.ca

皮尔区域——融合景观
www.peelregion.ca/watersmartpeel

多伦多市——天然草坪及花园养护
www.toronto.ca/pesticides

约克地区——水资源的未来
www.york.ca/waterfortomorrow

多伦多地区保护协会——健康的庭院资源
www.trca.on.ca/yards

加拿大野生动物保护联合会——热衷园艺
www.wildaboutgardening.org

常绿——本地植物数据库
www.nativeplants.evergreen.ca

去哪里买本土植物

建议尽可能地使用本土植物，可以在花园中心、苗圃和本地植物销售处买到。

北美原生植物协会——商业本土植物种植者
www.nanps.org

安大略入侵植物委员会——苗圃识别项目
www.ontarioinvasiveplants.ca

Downspout Disconnection

TRCA Sustainable Technologies Evaluation Program – Downspout Disconnection Fact Sheet
www.sustainabletechnologies.ca

Canadian Mortgage and Housing Corporation – Downspout Disconnection
www.cmhc-schl.gc.ca/en/inpr/su/waho/waho_014.cfm

City of Toronto – Downspout Disconnection Program
www.toronto.ca/downspout

Region of Peel – Downspout Disconnection Program
www.peelregion.ca/pw/water

Rain Barrels

Canadian Mortgage and Housing Corporation – Collecting and Using Rainwater at Home
www.cmhc-schl.gc.ca/en/co/grho/grho_016.cfm

Region of Peel – Rain Barrels
www.peelregion.ca/watersmartpeel

Rainbarrel.ca
www.rainbarrel.ca

Riversides Toronto Homeowners Guide to Rainfall – Rain Barrels
www.riversides.org/rainguide

断开的落水管

多伦多地区保护协会可持续技术评估项目——断开的落水管工作表
www. sustainabletechnologies. ca

加拿大抵押和住房公司——断开的落水管
www. cmhc - schl. ac. ca/en/inpr/su/waho/waho _ 014. cfm

多伦多市——落水管断开项目
www. toronto. ca/downspout

皮尔地区——落水管断开项目
www. peelregion. ca/pw/water

雨水桶

加拿大抵押和住房公司——收集和使用家庭雨水
www. cmhc - schl. gc. ca/en/co/grho/gtho 016 cfm

皮尔地区——雨水桶
www. peelregion. ca watersmartpeel

雨水桶
www. rainbarrel. ca

多伦多河边住宅雨水指南——雨水桶
www. riversides. org/rainguide

Rain Gardens

Canadian Mortgage and Housing Corporation – Rain Gardens
www.cmhc-schl.gc.ca/en/co/grho/grho_007.cfm

Green Venture – Rain Gardens
water.greenventure.ca/rain-gardens

TRCA Sustainable Technologies Evaluation Program – Bioretention Fact Sheet
www.sustainabletechnologies.ca

Green Communities Canada – RAIN garden tour
www.raingardentour.ca

Certified landscape professionals in your area
www.landscapeontario.com/findapro

Soakaways

Canadian Mortgage and Housing Corporation- Infiltration Trenches
www.cmhc-schl.gc.ca/en/inpr/su/waho/waho_013.cfm

TRCA Sustainable Technologies Evaluation Program – Soakaways, Infiltration Trenches and Chambers
www.sustainabletechnologies.ca

Innovative Stormwater Management Practices
www.iswm.ca

Government of British Columbia, Ministry of the Environment – Infiltration Practices
www.env.gov.bc.ca

雨水花园

加拿大抵押和住房公司——雨水花园
www. cmhc - schil. qc. ca/en/co/grho/grho 007. cfm

绿色冒险——雨花园水
www. water. greenventure. ca/rain-gardens

多伦多地区保护协会可持续技术评估项目——生物滞留池工作表
www. sustainabletechnologies. ca

加拿大绿色社区——雨水花园
www. raingardentour. ca

你所在地区的注册景观专业人士
www. landscapeapeontario. com/findapro

渗水坑

加拿大抵押和住房公司——渗透沟
www. cmhc - schl. ge. ca/en/inpr/su/waho/waho _ 013. cfm

多伦多地区保护协会可持续技术评估项目——渗水坑、渗透沟和渗透室
www. sustainabletechnolgies. ca

创新的雨水管理方法
www. iswm. ca

不列颠哥伦比亚省政府，环境部——渗透实践
www. env. gov. bc. ca

Permeable Pavement

TRCA Sustainable Technologies Evaluation Program – Permeable Pavement
www.sustainabletechnologies.ca

Green Communities Canada – Depave Paradise
www.depaveparadise.ca

IMPORTANT
CALL BEFORE YOU DIG!

ON1CALL
CALL BEFORE YOU DIG
It's The Law!
www.on1call.com
1-800-400-2255

透水铺装

多伦多及地区保护协会可持续技术评估项目——透水铺装
www. sustainabletechnologies. ca

加拿大绿色社区——透水铺装
www. depaveparadise. ca

重要提示

开挖前先打个电话！

ON1CALL

开挖前先打个电话！
这是法律！
www. on1call. com
1－800－400－2255

This Homeowners Guide to Stormwater Landscaping is intended to introduce community residents to projects that they can install on their own properties. It was created by Toronto and Region Conservation (TRCA) as part of a four year project in the Valleybrook Community along the Etobicoke Creek in Brampton, Ontario. It was made possible through funding support from the Ontario Trillium Foundation and the Region of Peel.

Region of Peel
Working for you

Ontario
Trillium Foundation

Fondation Trillium
de l'Ontario

Toronto and Region
Conservation
for The Living City

Member of Conservation Ontario

这本手册主要是让社区居民在自己的私有房屋和庭院中，建造和使用相应的雨水利用和雨水景观工程。这本手册是由多伦多及地区保护协会（TRCA）撰写的，也是布鲁克溪谷社区为期四年的项目的一部分，该社区位于安大略省布兰普顿的伊陶碧谷湾。它是通过安大略崔莉姆基金会和皮尔地区的资金支持才得以实现的。

Region of Peel
Working for you

Ontario Trillium Foundation
Fondation Trillium de l'Ontario

Toronto and Region Conservation for The Living City
Member of Conservation Ontario

国家示范性高等职业院校城市水利专业课程改革系列教材
高等职业教育水利类海绵城市特色教材

庭院雨水景观设计实用手册

多伦多及地区保护协会（TRCA） 编
冯峰 刘翠 靳晓颖 译

中国水利水电出版社
www.waterpub.com.cn
·北京·

内 容 提 要

本书为高职高专和成人高校、应用型本科的城市水利、生态水利、海绵城市、水务管理和水利工程等专业通用中英文对照教材。全书共分10个项目，包括引言、场地评估、土地绿化、断开的落水管、雨水桶、雨水花园、渗水坑、透水铺装、雨水的连通路径、相关资源等。

本书也可作为其他相近专业的教学参考书，也可供有关工程技术人员参考，还可供喜爱亲自动手设计庭院雨水景观的读者自学使用。

图书在版编目（CIP）数据

庭院雨水景观设计实用手册 / 多伦多及地区保护协会（TRCA）编 ; 冯峰，刘翠，靳晓颖译. -- 北京 : 中国水利水电出版社，2019.6

国家示范性高等职业院校城市水利专业课程改革系列教材 高等职业教育水利类海绵城市特色教材

ISBN 978-7-5170-7743-5

Ⅰ. ①庭… Ⅱ. ①多… ②冯… ③刘… ④靳… Ⅲ. ①城市水利－城市景观－景观设计－高等职业教育－教材 Ⅳ. ①TU-856

中国版本图书馆CIP数据核字(2019)第112862号

书　　名	国家示范性高等职业院校城市水利专业课程改革系列教材 高等职业教育水利类海绵城市特色教材 **庭院雨水景观设计实用手册** TINGYUAN YUSHUI JINGGUAN SHEJI SHIYONG SHOUCE
作　　者	多伦多及地区保护协会（TRCA）编 冯峰　刘翠　靳晓颖　译
出版发行	中国水利水电出版社 （北京市海淀区玉渊潭南路1号D座　100038） 网址：www. waterpub. com. cn E-mail：sales@waterpub. com. cn 电话：（010）68367658（营销中心）
经　　售	北京科水图书销售中心（零售） 电话：（010）88383994、63202643、68545874 全国各地新华书店和相关出版物销售网点
排　　版	中国水利水电出版社微机排版中心
印　　刷	北京博图彩色印刷有限公司
规　　格	285mm×210mm　横16开　7.5印张　267千字
版　　次	2019年6月第1版　2019年6月第1次印刷
印　　数	0001—1000册
定　　价	**49.00**元

前　言

这本书的诞生过程有些意思。

2017年8月16—25日，我和靳晓颖老师一同前往加拿大弗莱明学院进行交流访问。弗莱明学院可替代性污水处理中心（CAWT）的资深研究员Gordon Balch教授热情地接待了我们，并安排了充实的考察和调研内容，先后参观了弗莱明学院的地质实训室、可替代性污水处理中心，以及卡沃萨湖自然保护区、Otonabee流域保护管理局、饮用水净化厂等。

8月23日，在Otonabee流域保护管理局办公室，我们第一次看到*Greening Your Grounds*这本手册时，就感觉它特别实用，从场地测量、土壤判别、渗透率测定，到画图、选择方案、最终动手实施，全过程手把手地指导，不仅能锻炼动手能力，也能将思想、审美融入到庭院的雨水景观中。因此，我和靳晓颖老师商量着回国后把它翻译成中文，作为中英文对照的专业教材，让水利工程、城市水利专业的学生开拓眼界，在学习海绵城市、雨水景观工程、生态工程等相关专业知识的同时，提高专业英语的应用能力。

回国后，我们和团队的老师商量了一下，大家都认为这是个不错的主意。于是团队成员进行了分工，我负责引言、场地评估、土地绿化、断开的落水管、雨水的连通路径、相关资源部分，靳晓颖老师负责雨水桶、雨水花园部分，刘翠老师负责渗水坑、透水铺装部分。翻译工作进行的同时，我在Gordon Balch教授的帮助下，一直与多伦多及地区保护协会（TRCA）教育与培训机构的高级经理Joanne Jeffery沟通这本书的翻译版权问题，从2018年6月一直沟通到9月，近30封电子邮件来回协商后，Joanne Jeffery终于同意让我们免费使用这本书的翻译版权。

这本跨越了万里之远从加拿大来到中国的手册，前后经历了2年时间，终于即将问世。在这里特别感谢Gordon Balch、Joanne Jeffery给予的大力支持和帮助。本书在整个翻译和整理过程中，得到了祝玉华教授、焦爱萍教授、陶永霞教授的大力帮助和指导，在此表示衷心的谢意。翻译和整理过程中参考和借鉴了有关教材和科技文献资料的内容，在此一并表示感谢。

本书所有专业词汇的音标、释义均出自第9版《牛津高阶英汉双解词典》。由于译者水平有限，书中难免存在缺点和疏漏，恳请各位专家、同行和广大读者批评指正。

冯峰

2019年1月

Table of Contents

目　录

专业词汇

assessment/ə'sesmənt/*n*. 评估；估计

naturalization/ˌnætʃərəlaɪ'zeɪʃən/*n*. 归化，移入；移植

downspout/'daʊnspaʊt/*n*. 水落管，落水管（将雨水从屋顶排至水沟）

disconnection/ˌdɪskə'nekʃn/*n*. 断开；分离，切断

rain barrel 接雨水的桶，雨水桶

soakaway/'səʊkəˌweɪ/*n*. 渗水井，渗水坑

permeable/'pɜːmɪəbl/*adj*. （材料，膜）可渗透的，具有渗透性的

pavement/'peɪvmənt/*n*. 人行道，路面

connect/kə'nekt/*vt*. 连接；联结